CELLS
Building Blocks of Life

Anthea Maton
Former NSTA National Coordinator
Project Scope, Sequence, Coordination
Washington, DC

Jean Hopkins
Science Instructor and Department Chairperson
John H. Wood Middle School
San Antonio, Texas

Susan Johnson
Professor of Biology
Ball State University
Muncie, Indiana

David LaHart
Senior Instructor
Florida Solar Energy Center
Cape Canaveral, Florida

Maryanna Quon Warner
Science Instructor
Del Dios Middle School
Escondido, California

Jill D. Wright
Professor of Science Education
Director of International Field Programs
University of Pittsburgh
Pittsburgh, Pennsylvania

Prentice Hall
Englewood Cliffs, New Jersey
Needham, Massachusetts

Prentice Hall Science

Cells: Building Blocks of Life

Student Text and Annotated Teacher's Edition
Laboratory Manual
Teacher's Resource Package
Teacher's Desk Reference
Computer Test Bank
Teaching Transparencies
Product Testing Activities
Computer Courseware
Video and Interactive Video

The illustration on the cover, rendered by Keith Kasnot, shows the organelles that make up a typical cell.

Credits begin on page 121.

SECOND EDITION

© 1994, 1993 by Prentice-Hall, Inc., Englewood Cliffs, New Jersey 07632.

ISBN 0-13-400466-3

7 8 9 10   97 96 95

Prentice Hall
A Division of Simon & Schuster
Englewood Cliffs, New Jersey 07632

STAFF CREDITS

Editorial:	Harry Bakalian, Pamela E. Hirschfeld, Maureen Grassi, Robert P. Letendre, Elisa Mui Eiger, Lorraine Smith-Phelan, Christine A. Caputo
Design:	AnnMarie Roselli, Carmela Pereira, Susan Walrath, Leslie Osher, Art Soares
Production:	Suse F. Bell, Joan McCulley, Elizabeth Torjussen, Christina Burghard
Photo Research:	Libby Forsyth, Emily Rose, Martha Conway
Publishing Technology:	Andrew Grey Bommarito, Deborah Jones, Monduane Harris, Michael Colucci, Gregory Myers, Cleasta Wilburn
Marketing:	Andrew Socha, Victoria Willows
Pre-Press Production:	Laura Sanderson, Kathryn Dix, Denise Herckenrath
Manufacturing:	Rhett Conklin, Gertrude Szyferblatt

Consultants

Kathy French	National Science Consultant
Jeannie Dennard	National Science Consultant
Brenda Underwood	National Science Consultant
Janelle Conarton	National Science Consultant

Contributing Writers

Linda Densman
Science Instructor
Hurst, TX

Linda Grant
Former Science Instructor
Weatherford, TX

Heather Hirschfeld
Science Writer
Durham, NC

Marcia Mungenast
Science Writer
Upper Montclair, NJ

Michael Ross
Science Writer
New York City, NY

Content Reviewers

Dan Anthony
Science Mentor
Rialto, CA

John Barrow
Science Instructor
Pomona, CA

Leslie Bettencourt
Science Instructor
Harrisville, RI

Carol Bishop
Science Instructor
Palm Desert, CA

Dan Bohan
Science Instructor
Palm Desert, CA

Steve M. Carlson
Science Instructor
Milwaukie, OR

Larry Flammer
Science Instructor
San Jose, CA

Steve Ferguson
Science Instructor
Lee's Summit, MO

Robin Lee Harris Freedman
Science Instructor
Fort Bragg, CA

Edith H. Gladden
Former Science Instructor
Philadelphia, PA

Vernita Marie Graves
Science Instructor
Tenafly, NJ

Jack Grube
Science Instructor
San Jose, CA

Emiel Hamberlin
Science Instructor
Chicago, IL

Dwight Kertzman
Science Instructor
Tulsa, OK

Judy Kirschbaum
Science/Computer Instructor
Tenafly, NJ

Kenneth L. Krause
Science Instructor
Milwaukie, OR

Ernest W. Kuehl, Jr.
Science Instructor
Bayside, NY

Mary Grace Lopez
Science Instructor
Corpus Christi, TX

Warren Maggard
Science Instructor
PeWee Valley, KY

Della M. McCaughan
Science Instructor
Biloxi, MS

Stanley J. Mulak
Former Science Instructor
Jensen Beach, FL

Richard Myers
Science Instructor
Portland, OR

Carol Nathanson
Science Mentor
Riverside, CA

Sylvia Neivert
Former Science Instructor
San Diego, CA

Jarvis VNC Pahl
Science Instructor
Rialto, CA

Arlene Sackman
Science Instructor
Tulare, CA

Christine Schumacher
Science Instructor
Pikesville, MD

Suzanne Steinke
Science Instructor
Towson, MD

Len Svinth
Science Instructor/
Chairperson
Petaluma, CA

Elaine M. Tadros
Science Instructor
Palm Desert, CA

Joyce K. Walsh
Science Instructor
Midlothian, VA

Steve Weinberg
Science Instructor
West Hartford, CT

Charlene West, PhD
Director of Curriculum
Rialto, CA

John Westwater
Science Instructor
Medford, MA

Glenna Wilkoff
Science Instructor
Chesterfield, OH

Edee Norman Wiziecki
Science Instructor
Urbana, IL

Teacher Advisory Panel

Beverly Brown
Science Instructor
Livonia, MI

James Burg
Science Instructor
Cincinnati, OH

Karen M. Cannon
Science Instructor
San Diego, CA

John Eby
Science Instructor
Richmond, CA

Elsie M. Jones
Science Instructor
Marietta, GA

Michael Pierre McKereghan
Science Instructor
Denver, CO

Donald C. Pace, Sr.
Science Instructor
Reisterstown, MD

Carlos Francisco Sainz
Science Instructor
National City, CA

William Reed
Science Instructor
Indianapolis, IN

Multicultural Consultant

Steven J. Rakow
Associate Professor
University of Houston—
Clear Lake
Houston, TX

English as a Second Language (ESL) Consultants

Jaime Morales
Bilingual Coordinator
Huntington Park, CA

Pat Hollis Smith
Former ESL Instructor
Beaumont, TX

Reading Consultant

Larry Swinburne
Director
Swinburne Readability
Laboratory

CONTENTS

CELLS: BUILDING BLOCKS OF LIFE

Activity Bank/Reference Section

Features

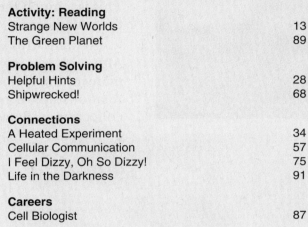

CONCEPT MAPPING

Throughout your study of science, you will learn a variety of terms, facts, figures, and concepts. Each new topic you encounter will provide its own collection of words and ideas—which, at times, you may think seem endless. But each of the ideas within a particular topic is related in some way to the others. No concept in science is isolated. Thus it will help you to understand the topic if you see the whole picture; that is, the interconnectedness of all the individual terms and ideas. This is a much more effective and satisfying way of learning than memorizing separate facts.

Actually, this should be a rather familiar process for you. Although you may not think about it in this way, you analyze many of the elements in your daily life by looking for relationships or connections. For example, when you look at a collection of flowers, you may divide them into groups: roses, carnations, and daisies. You may then associate colors with these flowers: red, pink, and white. The general topic is flowers. The subtopic is types of flowers. And the colors are specific terms that describe flowers. A topic makes more sense and is more easily understood if you understand how it is broken down into individual ideas and how these ideas are related to one another and to the entire topic.

It is often helpful to organize information visually so that you can see how it all fits together. One technique for describing related ideas is called a **concept map**. In a concept map, an idea is represented by a word or phrase enclosed in a box. There are several ideas in any concept map. A connection between two ideas is made with a line. A word or two that describes the connection is written on or near the line. The general topic is located at the top of the map. That topic is then broken down into subtopics, or more specific ideas, by branching lines. The most specific topics are located at the bottom of the map.

To construct a concept map, first identify the important ideas or key terms in the chapter or section. Do not try to include too much information. Use your judgment as to what is really important. Write the general topic at the top of your map. Let's use an example to help illustrate this process. Suppose you decide that the key terms in a section you are reading are School, Living Things, Language Arts, Subtraction, Grammar, Mathematics, Experiments, Papers, Science, Addition, Novels. The general topic is School. Write and enclose this word in a box at the top of your map.

SCHOOL

Now choose the subtopics—Language Arts, Science, Mathematics. Figure out how they are related to the topic. Add these words to your map. Continue this procedure until you have included all the important ideas and terms. Then use lines to make the appropriate connections between ideas and terms. Don't forget to write a word or two on or near the connecting line to describe the nature of the connection.

Do not be concerned if you have to redraw your map (perhaps several times!) before you show all the important connections clearly. If, for example, you write papers for Science as well as for Language Arts, you may want to place these two subjects next to each other so that the lines do not overlap.

One more thing you should know about concept mapping: Concepts can be correctly mapped in many different ways. In fact, it is unlikely that any two people will draw identical concept maps for a complex topic. Thus there is no one correct concept map for any topic! Even

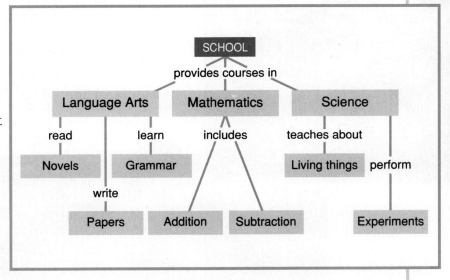

though your concept map may not match those of your classmates, it will be correct as long as it shows the most important concepts and the clear relationships among them. Your concept map will also be correct if it has meaning to you and if it helps you understand the material you are reading. A concept map should be so clear that if some of the terms are erased, the missing terms could easily be filled in by following the logic of the concept map.

CELLS

Building Blocks of Life

From the smallest bacteria to the largest whales, the building blocks of all living things are cells.

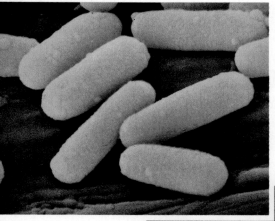

What do you think of when you hear the words building blocks? Perhaps your mind drifts back to a younger age when you may have spent a good deal of time playing with wooden or plastic building blocks. (If you are like the author of this textbook, you still enjoy playing with them.) Or perhaps you think of the brick, steel, and wood out of which country houses and modern skyscrapers are constructed. If you enjoy the physical sciences, you might even think of atoms—the building blocks of matter.

In this textbook, you are going to explore very special building blocks—the building blocks of life! What are the building blocks of

CHAPTERS

life? You can probably guess from the title of this textbook that they are cells. All living things—from the tiniest bacteria to the largest blue whales—are made of microscopic building blocks called cells. Cells are the framework upon which living things are constructed. Why are cells so special? Unlike all other kinds of building blocks, cells are alive!

What exactly are cells? How do they perform the activities necessary for life? Are all cells the same or do they vary from organism to organism? These are just a few of the questions you will be able to answer when you have finished your exploration of one of nature's greatest accomplishments—the living cell.

▲ Like these children in a tug of war, your cells must work together as a team to keep your body running smoothly.

Discovery Activity

How Long Is Long? How Small Is Small?

As you read this textbook, you will encounter some rather large numbers and some rather small numbers. For example, you will discover that the history of the Earth and the evolution of cells is a story that began about 4.5 billion years ago. A billion is a 1 followed by 9 zeros! You will also discover that the smallest cells measure about 0.2 micrometer in height. A micrometer is a millionth of a meter, or a fraction with 1 as the numerator and 1 million as the denominator (1/1,000,000).

1. Construct a time line starting with the formation of the Earth and ending with the present day. Include any historical dates you feel are important. As you read this textbook, add new dates to your time line.

2. The average student is about 1.5 meters tall. If you began stacking up the smallest cells, how many would you have to add before your stack reached 1.5 meters?

   ■ Do you now have a better idea of how long is long and how small is small?

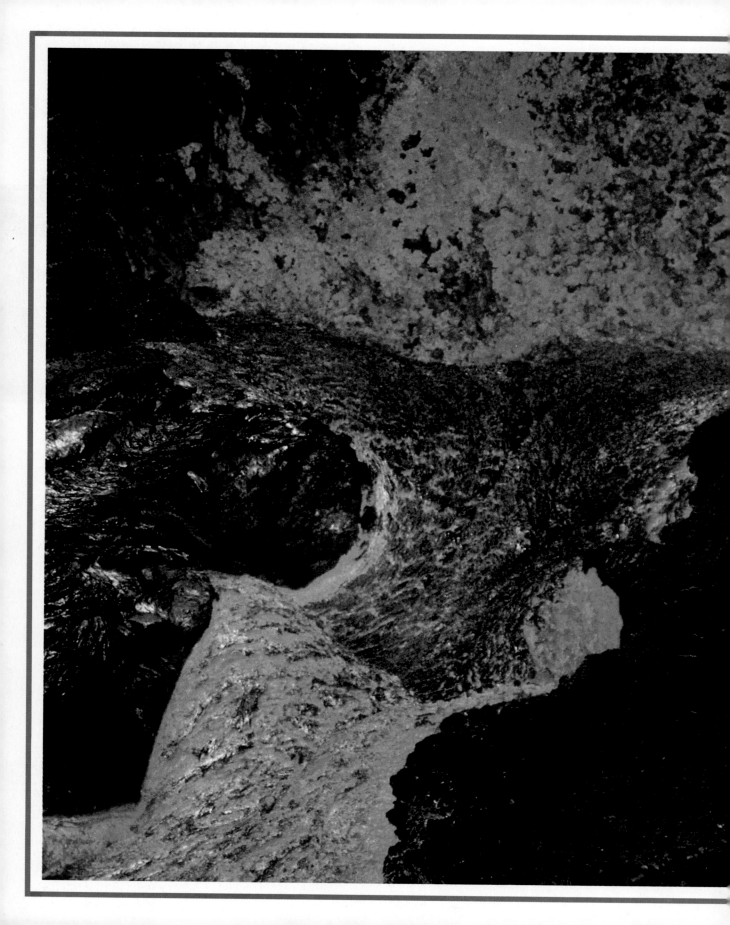

The Nature of Life

A huge cloud of swirling dust and hot gases glows in the eerie darkness of space. Over many billions of years, gravity begins to pull the dust and gas toward the center of the cloud. In time, the cloud condenses into a giant sphere of matter. A planet has formed.

Half a billion years pass. The planet has begun to cool, and solid rocks dot its surface. Volcanoes spring up everywhere, shaking the planet with their constant eruptions. A poisonous atmosphere begins to form.

Another 200 million years pass. The planet is now cool enough to allow liquid water to flow on its surface. Thunderstorms begin to drench the planet with rain—year after year after year. In time, planet-wide oceans form. Although it will take another 300 million years, eventually living things will call these oceans home. Slowly but surely, these living things become more complex and begin to change, or evolve. Over the next 3.5 billion years, many living things will come and go on this planet as it floats in space. But one day very special living things will arise—living things that can pick up this textbook and discover what life is all about. Try it!

Journal *Activity*

On Your Own When you read about the origin of Earth and the beginning of life, you read about events occurring over millions, even billions, of years. To help you get a better sense of time relationships, in your journal begin your time line with the major events discussed in this chapter. You may add any other events you desire—perhaps your birthday.

About 4 billion years ago, spectacular volcanic eruptions on Earth were a common occurrence.

Guide for Reading

Focus on these questions as you read.

▶ How old is Earth, and when did life first form on it?

▶ How did the development of photosynthesis pave the way for the existence of modern living things?

1–1 The Origin of Life

Slowly, the scientist fills the clear glass flask. First he pours in three colorless gases. The odor is awful, stinging the scientist's nose and bringing tears to his eyes. Then the scientist adds another gas. Nothing seems to happen. The flask looks empty. But if everything goes right, the gases it contains may be changed into something very special!

The mixture needs a spark to produce the necessary change. The scientist sends a surge of electricity through the flask again and again. At the same time, he shines ultraviolet light at the flask. Suddenly, a sticky brown coating begins to form on the walls of the flask. The mixture of gases inside is changing—turning into substances that may help to solve the key mystery of life.

Magic? It may seem to be, and at times the scientist may seem to be a magician. But his exciting experiment was actually performed, and its results are being used by scientists today as they attempt to study the "stuff of life."

In 1953, the American scientist Stanley Miller mixed together three gases: hydrogen, methane, and ammonia. To this mixture he added gaseous water. Then he passed an electric current to simulate lightning through the colorless mixture. Soon a brown tarlike substance streaked the sides of the container. Dr. Miller analyzed the tarlike substance and found that it contained several of the same substances that make up all living things. From nonliving chemicals, Stanley Miller had made some of the building blocks of life!

How can lifeless chemicals change into the matter that makes up life? To answer that question, we must go back to the formation of Earth.

The Early Earth

You have read about the formation of Earth at the beginning of this chapter. Naturally, no one was there to see Earth form and to record exactly how it happened. Although we lack direct evidence about early Earth, we can make some basic generalizations about it. But keep in mind that much is still unclear

Figure 1–1 *In 1953, Stanley Miller demonstrated that some of the chemicals that make up living things could have formed on ancient Earth.*

Figure 1–2 *You can see from this illustration of ancient Earth that most modern living things would not easily survive those rugged conditions.*

and that the complete picture has not yet been achieved.

Planet Earth formed about 4.6 billion years ago. (A billion is a 1 followed by 9 zeros!) But it would be more than half a billion years before the planet cooled and a rocky surface was created. And, as you have read, hundreds of millions of years more would pass before the oceans formed.

If you could turn back time and visit ancient Earth, you would be in for quite a surprise. The atmosphere was quite poisonous and could not support life as we know it. Scientists do not know the exact composition of that early atmosphere. Most agree that it contained some water vapor (gaseous water), carbon monoxide and carbon dioxide, nitrogen, hydrogen sulfide, methane, and hydrogen cyanide. Did the atmosphere contain any oxygen, so necessary for life on Earth today? The answer is unclear. Many scientists believe there was little or no oxygen in the early atmosphere. But recent evidence indicates such theories may have to be adjusted; the early atmosphere may have contained oxygen. In either case, you would not have been able to breathe the air and survive.

ACTIVITY READING

Strange New Worlds

If you like science fiction and enjoy reading about new worlds, you might like to read *Last and First Men* by Olaf Stapledon. In his journey into the far future, Stapledon writes of the evolution of people and the many worlds they conquer.

The Molecules of Life Form

The experiment performed by Stanley Miller showed that compounds necessary for life can be produced from nonliving substances. Miller's experiment opened the door to the exploration of how life formed on Earth. Today we know that the gases Miller placed in the flask do not match the atmosphere of early Earth. So we cannot say that life began in a manner similar to Miller's experiment. But we can say that the chemicals that make up living things could have been produced on early Earth. The exact process that would have made this possible is another question scientists have not as yet answered.

A batch of chemicals is a long way from a living thing. And just as scientists cannot say with certainty how the chemicals of life formed, they also cannot say how these chemicals came together to form the first living things. But based on current evidence, they can develop theories.

One theory begins with the notion that the early oceans began somehow to fill with the chemicals that make up living things. We can think of these oceans as a kind of "soup" containing the substances needed for life. Using solid evidence about the behavior of matter, scientists do have ideas about how some chemicals in the soup may have come together to form the beginning of life. For example, we now know that under the right conditions amino acids will link together on their own to form small chains. This is significant because a chain of amino acids is called a protein and proteins are among the basic building blocks of living things. Other chemicals, which scientists believe might have been in that original soup, have also been observed to link together on their own to produce carbohydrates, alcohols, and fatty substances called lipids. (You will learn more about proteins, carbohydrates, lipids, and other chemicals of life later in this chapter.) So

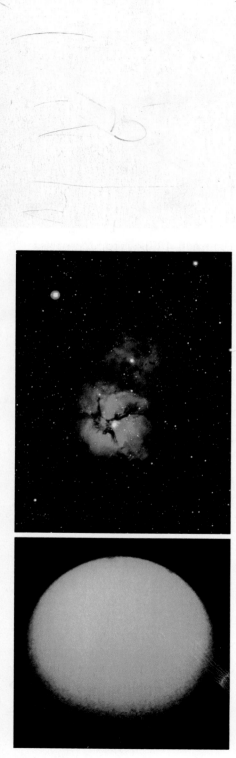

Figure 1–3 *Analysis of distant nebula has shown that some chemicals of life form spontaneously in the dust and gas of outer space (top). Many scientists feel that conditions in the atmosphere of Titan, Saturn's largest moon, may be similar to conditions on primitive Earth (bottom). Some scientists suggest that microscopic living things may have evolved on Titan.*

Figure 1–4 *One theory proposes that much of the water on Earth was carried to the primitive planet as frozen ice in meteors and asteroids.*

although we cannot say exactly how it happened, we can say with some certainty that it was possible for chemicals in the soup to produce some of the building blocks of life.

How Cells Formed

Later in this chapter you will discover that one of the basic characteristics of living things is that they are made of cells. So it is safe to assume that life on Earth began when the first cells formed. How did that process occur? Again, there are no clear answers—only theories based on current evidence.

One theory states that the first cells arose in shallow pools containing the early soup of chemicals. These cells formed as the chemicals in the soup collected into droplets. The droplets were surrounded by a wall, or barrier, that kept the chemicals inside and the soup outside. Over time, the substances necessary for life developed and the droplets became true cells.

Yet another theory states that the first cells formed in beds of clay on early Earth. The structure of the clay allowed the chemicals necessary for life to become concentrated, or trapped in the clay. As time went on, these chemicals grouped together to produce the first living cells.

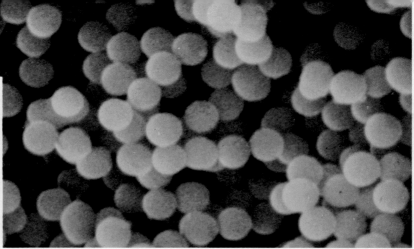

Figure 1–5 *These droplets, magnified 3000 times, were created in the laboratory of Sidney Fox. Although the droplets are not alive, they can perform some of the basic life functions. Some droplets can actually reproduce by dividing into two separate droplets.*

There are other theories regarding the origin of the first true cells. At this time, however, no one can say whether any one theory is right or wrong. What can be said is that somehow, through some process, the chemicals that make up living things did group together and form the first cells. Once these first cells formed, the parade of life on Earth began—a parade that continues to this day.

The First True Cells

Scientists have discovered fossils (remains of living things) that indicate that the first true cells evolved and inhabited Earth as far back as 3.5 billion years ago. The origin of these first true cells is still unknown. But no matter how they formed, some generalizations about them can be made.

The first true cells were doubtless organisms that did not require oxygen. Remember, the early atmosphere probably contained little or no oxygen. So it is safe to assume that early cells did not require oxygen. It is also safe to assume that these early cells were consumers, or organisms that did not produce their own food. The soup in which these early forms of life floated was their source of food. And indeed, the early cells had plenty of food to feed upon. But as you might expect, over time the food in the soup gradually began to dwindle. In order for life to have continued, cells capable of producing their own food must have evolved. These cells used chemicals from their environment to produce food and energy. Thus, living things began to change from consumers to producers.

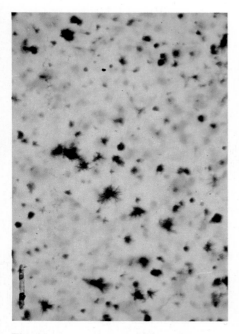

Figure 1–6 *You can see 2-billion-year-old fossils of bacteria cells in this thin slice of rock. When do scientists estimate the first true cells formed?*

Photosynthesis

Earth today is filled with both consumers and producers. You, for example, are a consumer. You must eat food in order to obtain energy and important nutrients. Green plants, on the other hand, are producers. They use chemicals in their environment and the energy of sunlight to produce their own food. Green plants produce food in a process called photosynthesis. (You will learn much more about photosynthesis in Chapter 4.)

Scientists theorize that at some point certain early cells were able to perform photosynthesis. This is an extremely important event in the history of Earth. Do you know why? One of the waste materials plants produce during photosynthesis is a gas called oxygen. As early cells began to perform photosynthesis, they also began to change the atmosphere of Earth. Over millions of years, Earth's poisonous atmosphere was changed to one that contains about one fifth part oxygen. The stage was now set for organisms that use oxygen to evolve.

Modern Cells Form

The production of oxygen by early cells transformed Earth and gave rise to cells that use oxygen for many of the chemical reactions that take place within them. Cells that use oxygen are much more efficient in their production of energy than cells that do not use oxygen. With greater energy efficiency, cells were free to evolve in a wide variety of ways.

The history of cell evolution is a fascinating topic and one that is hotly debated. How did cells that use oxygen first evolve? How were the cells we find in living things today produced? These questions are as yet unanswered. Perhaps they never will be. But one thing is certain: Along with photosynthesis and cells that use oxygen came other evolutionary advances.

One advance was the appearance of multicellular organisms, or organisms that contain many cells. The jump from single-celled (unicellular) organisms to multicellular organisms was an important leap in the development of living things. In time, cells in multicellular organisms began to specialize and perform

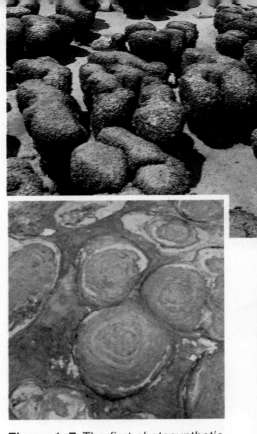

Figure 1–7 *The first photosynthetic organisms grew in layered mats called stromatolites. Shark Bay, Australia, is one of the few places where living stromatolites still exist. Fossils of stromatolites can be found throughout the world.*

ACTIVITY
DISCOVERING

A Different Light

1. Plant several of the same type of seeds in five different containers.

2. When the seedlings are about 2.5 cm above the soil, cover each top of the container with one of the following cellophane sheets: red, green, yellow, blue, and clear.

How do the different sheets of cellophane affect plant growth?

Figure 1–8 *In asexual reproduction, such as the division of a protozoan, each new cell is an exact copy of the original cell. In sexual reproduction, however, offspring inherit traits from two parents, increasing variation in the species.*

specific functions. A new form of reproduction developed as well. Early cells reproduced by splitting into two cells. And in time a type of reproduction called sexual reproduction developed. During sexual reproduction, sex cells from two different parents join together to form a new organism. With the development of multicellular organisms and sexual reproduction, all the facets that make up life had arisen. All that was left was to wait and see what evolved—a process we are still exploring today.

1–1 Section Review

1. What is the age of Earth?
2. How do we know that the first living things did not require oxygen?
3. Are you unicellular or multicellular? Explain.

Critical Thinking—*Drawing Conclusions*
4. Could the events that led to the first true cells recur on Earth today? Explain your answer.

Guide for Reading

Focus on this question as you read.

▶ *What are the characteristics of living things?*

1–2 Characteristics of Living Things

Take a short walk in the city or the country and you will see an enormous variety of living things. In fact, scientists estimate that there are up to 10 million different types of organisms on Earth, ranging in size from single-celled bacteria to huge blue whales. Yet all these living things are composed mainly of the same basic elements: carbon, hydrogen, nitrogen, and oxygen. These elements make up the gases Stanley Miller placed in his flask. And often, along with iron, calcium, phosphorus, and sulfur, these four elements link together in chains, rings, and loops to form the stuff of life.

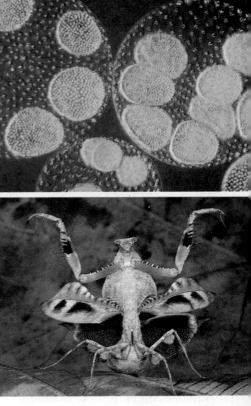

Figure 1–9 *The walrus (left), bottle tree (center),* Volvox *(top right), and dead-leaf mantis (bottom right) are among the great variety of living things that inhabit planet Earth. Yet despite their diversity, all living things are made of the same basic elements.*

Well-known chemical rules govern the way these elements combine and interact. But less well understood is what gives this collection of chemicals a very special property—the property of life.

Spontaneous Generation

People did not always understand that living matter is so special. Until the 1600s, most people believed in **spontaneous generation.** According to this theory, life could spring from nonliving matter. For example, people believed that mice came from straw and that frogs and turtles developed from rotting wood and mud at the bottom of a pond.

In 1668, an Italian doctor named Francesco Redi disproved the spontaneous-generation theory. Here is how he did it. In those days, maggots (a wormlike stage in the life cycle of a fly) often appeared on decaying (rotting) meat. People believed that the rotten meat had actually turned into maggots. This could only mean that flies (adult maggots) formed from dead animals (meat)—or that nonliving things could give rise to living things. In a series of

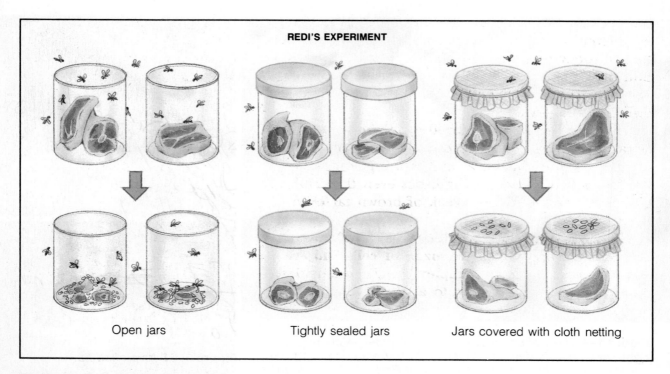

REDI'S EXPERIMENT

Open jars | Tightly sealed jars | Jars covered with cloth netting

Figure 1–10 *In Redi's experiment, no maggots were found on the meat in jars covered with netting or in those tightly sealed. Maggots appeared on the meat only when flies were able to enter the jars and lay eggs.*

ACTIVITY

WRITING

Life From Life

Using reference materials in the library, find out about each of these scientists:

John Needham
Lazzaro Spallanzani
Louis Pasteur

In a written report, describe how each scientist used the scientific method to prove or disprove the theory of spontaneous generation.

experiments, which are illustrated in Figure 1–10, Redi proved that the maggots hatched from eggs laid by flies.

Redi's experiment was quite simple. He placed rotting meat in several jars. He left two jars open and sealed two others. He covered the third set of jars with a cloth netting. The netting let in air but did not allow flies to land on the meat. In a few days, Redi observed flies on and above the meat in the open jars. There were no flies in the sealed jars or in the jars covered with netting. When people attempted to discredit Redi's results by claiming no flies were created in the meat in the sealed jars because the jars kept out air, Redi pointed out that air could enter the jars covered with netting. In a simple but elegant experiment, Redi had proved that spontaneous generation does not occur. Today there is no doubt that living things can arise only from other living things.

At this chapter's beginning, you read about the origin of life. Now you have read about spontaneous generation and how it is an incorrect theory. Is this double talk? Could life have arisen from nonliving things on early Earth, even though it does not occur on Earth today? The answer is yes. The conditions on early Earth were such that living things could arise from the soup of chemicals that formed on the Earth. Today that soup no longer exists. The formation of life as it occurred on early Earth cannot occur on its own again—at least not on Earth. On other planets—who knows!

Now that we have explored a bit about how life began (and why those processes cannot recur), it is time to examine just what makes living things so special. That is, what distinguishes even the smallest organism from a lifeless streak of brown tar on a laboratory flask?

There are certain characteristics that all forms of life share. Living things are made of cells and are able to move, perform complex chemical activities, grow and develop, respond to a stimulus, and reproduce.

Living Things Are Made of Cells

All living things are made of small units called cells. That is, cells are the basic building blocks of living things just as atoms are the basic building blocks of matter. Each cell contains living material surrounded by a border, or barrier, that separates the cell from its environment.

As Francesco Redi showed, cells are never formed by nonliving things. Cells come only from other cells. Nonliving matter may contain the remains of once-living cells, however. For example, firewood is made largely of cells that were once part of living trees.

Figure 1–11 *If you had no knowledge of trees, could you tell by a walk through a winter forest if these trees were living or nonliving?*

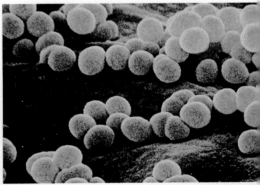

Figure 1–12 *All life on Earth is cellular. Some living things, such as these spherical bacteria, are unicellular. Other living things, such as the rhinoceros and her young, are multicellular.*

Some living things contain only a single cell. Single-celled, or unicellular, organisms include microscopic bacteria. (The Latin prefix *uni-* means one, so unicellular means single-celled.) The single cell in a unicellular organism can perform all the functions necessary for life. Most of the living things you are familiar with, such as cats and flowering plants, contain many cells, or are multicellular. What do you think the prefix *multi-* means in Latin?

Multicellular organisms may contain hundreds, thousands, or even trillions of cells. It has been estimated that humans contain about 6 trillion cells. Although the cells of multicellular organisms perform the basic functions of life, they are often specialized to perform a specific function in the organism. You will learn more about cells in Chapter 2. For now, all you need to remember is that the cell is the basic building block of living things.

Living Things Can Move

The ability to move through the environment is an important characteristic of many living things. Why? Animals must be able to move in order to find food and shelter. In times of danger, swift movement can be the difference between safety and death. Of course, animals move in a great many ways. Fins enable fish such as salmon to swim hundreds of kilometers in search of a place to mate. The kangaroo uses its entire body as a giant pogo stick to bounce along the Australian plains looking for scarce patches of grass upon which to graze. You have to move to turn the page and continue reading this chapter.

Figure 1–13 *In order to find food and shelter, most animals must be able to move. The arctic tern holds the long-distance record for birds in flight, as it covers nearly 32,000 kilometers in its yearly migration from the Arctic to the Antarctic and back. The kangaroo uses its body as a pogo stick as it bounces along the Australian plains. The larval crab gets around by hitching a ride on a jellyfish.*

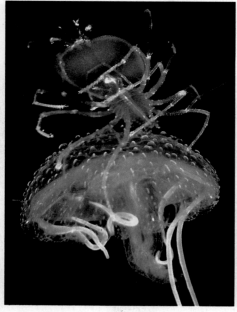

Most plants do not move in the same way animals do. Only parts of the plants move. The stems of many plants, for example, bend toward sunlight so the leaves on the plants can catch the sun's rays.

Living Things Perform Complex Chemical Activities

Building up and breaking down is a good way to describe the chemical activities that are essential to life. During some of these activities, simple substances combine to form complex substances. These substances are needed by an organism to grow, store energy, and repair or replace cells and other body parts. During other activities, complex substances are broken down, releasing energy and usable food substances. Together, these chemical activities are called **metabolism** (muh-TAB-uh-lih-zuhm). Metabolism is another characteristic of living things.

Metabolism is the sum total of all the chemical reactions that occur in a living thing. But before metabolism can begin, most organisms must perform a physical activity—taking in food.

INGESTION   All living things must either take in food or produce their own food. For most animals, **ingestion,** or eating, is as simple as putting food into their mouths.

Green plants do not have to ingest food. Green plants are able to make their own food. Using their roots, green plants absorb water and minerals from the soil. Tiny openings in the underside of their

Stimulus-Response Reactions

1. Hold your hands close to a friend's face. Quickly clap your hands while observing your friend's eyes.

2. While standing in front of a mirror, cover one of your eyes with your hand for a minute. Remove your hand and immediately look into the mirror and note any changes in your eye.

3. With a knife, cut a slice of lemon. **Caution**: *Be careful when using a knife.* Bring the lemon slice close to your mouth or put it in your mouth.

In a data table, describe the stimulus and the response for each of these activities.

Figure 1–14 *Green plants can make their own food, but animals must eat food. The grasshopper is feeding on a sunflower plant. The gecko, however, prefers a meatier dinner.*

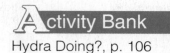
leaves allow carbon dioxide to enter. The green plants use the water and carbon dioxide, along with energy from the sun, to make food in the process called photosynthesis.

DIGESTION   Getting food into the body is a first step. Now the process of metabolism can begin. But there is a lot more to metabolism than just eating. The food must be digested in order to be used. **Digestion** is the process by which food is broken down into simpler substances. Later some of these simpler substances are reassembled into more complex materials for use in the growth and repair of the living thing. Other simple substances store energy that the organism will use for its many activities.

RESPIRATION   All living things require energy to survive. To obtain energy, living things combine oxygen with the products of digestion (in animals) or the products of photosynthesis (in green plants). The energy is used to do the work of the organism. The process by which living things take in oxygen and use it to produce energy is called **respiration.** You get the energy you need by combining the foods you eat with the oxygen you breathe.

EXCRETION   Not all the products of digestion and respiration can be used by an organism. Some products are waste materials that must be released. The process of getting rid of waste materials is called **excretion.** Like ingestion, excretion is a physical process. Without excretion, the waste products of digestion and respiration will build up in the body and eventually poison the organism.

Living Things Grow and Develop

The concept that living things grow is certainly not new to you. In fact, at this moment you are in the process of growth yourself. (How many times have you been told "When you grow up you can . . ."?)

When you think of growth, you probably think of something getting bigger. And that is certainly one part of growth. But growth can mean more than just an increase in size. Living things also become more complex, or develop, during the growth process. Sometimes this development results in dramatic changes. A tadpole, for example, swims for weeks in

Figure 1–15 *The gazelle obtains the energy it needs to run away from a predator by combining oxygen with food in a process called respiration. Although the killer whale spends most of its time under water, it must return to the surface to take in the oxygen it requires for respiration. Where do fish obtain the oxygen they require?*

Figure 1–16 *All living things grow and develop. Usually growth means simply getting larger, not changing form. But that is not always the case. This caterpillar will grow and develop into an adult lime butterfly.*

a summer pond. Then one day that tadpole becomes the frog that sits near the water's edge. And surely the caterpillar creeping through a garden gives little hint of the beautiful butterfly it will soon become. So both growth and development must be added to the list of characteristics of living things.

One of the important aspects of growth and development is **life span**. Life span is the maximum length of time a particular organism can be expected to live. Life span varies greatly from one type of organism to another. For example, an Indian elephant may live to be 80 years old. A bristlecone pine tree may live to be 5500 years old!

Living Things Respond to Their Environment

Scientists call each of the signals to which an organism reacts a **stimulus** (plural: stimuli). A stimulus is any change in the environment, or surroundings, of an organism that produces a **response** by that organism. A response is some action, movement, or change in behavior of the organism.

Figure 1–17 *In certain organisms, growth and development take up most of the life span. The mayfly spends two years in lakes, growing and developing into an adult. The adult, however, lives for only one day, during which it finds a mate, reproduces, and then dies. The life span of the bristlecone pine, on the other hand, can last up to 5500 years.*

Figure 1–18 *Living things respond to stimuli from their environment. What stimuli is the bat responding to? What will be the response of the frog?*

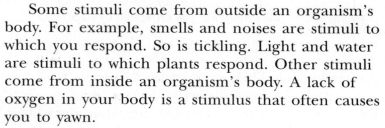

Some stimuli come from outside an organism's body. For example, smells and noises are stimuli to which you respond. So is tickling. Light and water are stimuli to which plants respond. Other stimuli come from inside an organism's body. A lack of oxygen in your body is a stimulus that often causes you to yawn.

Some plants have special responses that protect them. For example, when a gypsy moth caterpillar chews on a leaf of an oak tree, the tree responds by producing bad-tasting chemicals in its other leaves. The chemicals discourage all but the hungriest caterpillars from eating these leaves. Can you think of responses that help you protect yourself?

Living Things Reproduce

You probably know that dinosaurs lived millions of years ago and are now extinct. Yet crocodiles, which appeared on Earth before the dinosaurs, still exist today. An organism becomes extinct when it no longer produces other organisms of the same kind. In other words, all living things of a given kind would become extinct if they did not reproduce.

The process by which living things give rise to the same type of living thing is called reproduction. Crocodiles, for example, do not produce dinosaurs; crocodiles produce only more crocodiles. You are a human—not a water buffalo, duck, or tomato plant—because your parents are humans. An easy way to remember this is *like produces like.*

There are two different types of reproduction: **sexual reproduction** and **asexual reproduction**. Sexual reproduction usually requires two parents. Most

ACTIVITY

DOING

The Great Redi Experiment

1. Obtain 3 wide-mouthed jars. In each jar, place a piece of raw meat about the size of a half dollar.

2. Cover one jar with plastic wrap and another with two thicknesses of cheesecloth. Use rubber bands to hold the plastic wrap and cheesecloth in place. Leave the third jar uncovered.

3. Put the jars in a warm sunny place outdoors where they will remain undisturbed for 3 days. *Do not merely place the jars on a windowsill.*

4. After 3 days, examine the meat in each jar.

In which jar did you find maggots (young flies that resemble worms)? Did you find eggs in or on any of the jars? What does this activity tell you about spontaneous generation?

Figure 1–19 *The process by which living things give rise to the same type of living things is called reproduction. Does the bison reproduce through sexual or asexual reproduction?*

multicellular forms of plants and animals reproduce sexually.

Some living things reproduce from only one parent. This is asexual reproduction. When an organism divides into two parts, it is reproducing asexually. Bacteria reproduce this way. Yeast forms growths called buds, which break off and then form new yeast plants. Geraniums and African violets grow new plants from part of a stem, root, or leaf of the parent plant. All these examples illustrate asexual reproduction.

Sexual and asexual reproduction have an important function in common. In each case, the offspring receive a set of special chemical "blueprints," or plans. These blueprints determine the characteristics of that living thing and are passed from one generation to the next.

1–2 Section Review

1. List and describe the characteristics of living things. Which of these characteristics is not necessary for the survival of an organism?
2. Describe Redi's experiment on the spontaneous-generation theory.
3. Define metabolism. What are the main parts of metabolism?
4. Compare sexual and asexual reproduction.

Critical Thinking—*Applying Concepts*
5. A snowball rolling over fresh snow will grow larger. Explain why a rolling snowball is not a living thing.

ACTIVITY DISCOVERING

Living or Nonliving?

1. Obtain 6 mL of gelatin solution and 4 mL of gum arabic solution from your teacher. Add these solutions together in a test tube.

2. Stopper the test tube. Gently turn it upside down several times to mix the two solutions. **Note:** *Do not shake the test tube.*

3. Remove the stopper from the test tube and add 3 drops of weak hydrochloric acid. **CAUTION:** *Be careful when using acid.*

4. Dip a glass rod into the mixture. Touch a drop of the mixture onto a piece of pH paper. Compare the color of your pH paper to the color scale on the package of pH paper. Repeat steps 3 and 4 until the mixture reaches a pH of 4.

5. Place 2 drops of the mixture on a glass slide. Cover the slide with a coverslip and examine it under low power. Record your observations.

■ In what ways do the droplets seem to be living? In what ways do they seem to be nonliving?

Helpful Hints

You have been selected the new editor for your school newspaper. Your first assignment is the Helpful Hints column. Answer the following questions submitted by your fellow students. Your answers should relate to the fact that stimulus-response reactions are a method of protection against potentially harmful situations.

Relating Cause and Effect

1. Why do I squint in bright sunlight?

2. Why do birds migrate south in the winter?

3. When dirt gets in my eye, my eye blinks and produces tears. Why?

4. My dog pants heavily after a long run. Is she ill?

Guide for Reading

Focus on these questions as you read.

▶ *What are the basic needs of living things?*

▶ *What is the meaning of the term homeostasis?*

1–3 Needs of Living Things

Living things interact with one another as well as with their environment. These interactions are as varied as the living things themselves. Birds, for example, use dead twigs to build nests, but they eat live worms and insects. Crayfish build their homes in the sand or mud of streams and swamps. They absorb a chemical called lime from these waters and use it to build a hard body covering. Crayfish rely, however, on living snails and tadpoles for their food. And some people rely on crayfish for a tasty meal!

Figure 1–20 *All living things interact with their environment, as this hermit crab demonstrates. It crawls slowly up a tree while wearing the discarded shell of a snail.*

Clearly, living things depend on both the living and nonliving parts of their environment. **In order for a living organism to survive, it needs energy, food, water, oxygen, living space, and the ability to maintain a fairly constant body temperature.**

Energy

All living things need energy. The energy can be used in different ways depending on the organism. A lion uses energy to chase and capture its prey. The electric eel defends itself by shocking its attackers with electric energy.

What is the source of energy so necessary to living things? The primary source of energy for most living things is the sun. Does that surprise you? Plants use the sun's light energy to make food. Some animals feed on plants and in that way obtain the energy stored in the plants. Other animals then eat the plant eaters. In this way, the energy from the sun is passed on from one living thing to another. So next time you are eating a delicious meal— whether tacos, spaghetti, or bean-sprout salad—give silent thanks to the sun.

Food and Water

Food is a need of all living things. It is a source of energy as well as a supplier of the raw materials needed for growth, development, and repair of body parts.

ACTIVITY
DISCOVERING

Love That Light

Obtain two coleus plants of equal size. Using the two plants and a sunny window, design and perform an experiment that will show if plants move in response to light. *Hint:* You are using two plants for a reason.

■ What conclusions can you draw about the movement of plants in response to light?

Figure 1–21 *Plants, which make their own food through photosynthesis, are producers. Animals cannot make their own food. What term is given to organisms that cannot make their own food?*

Figure 1–22 *Not all plants rely only on photosynthesis for food. When a fly touches the tiny hairs lining a leaf of a Venus' flytrap plant, the leaf responds by closing and trapping the fly. The plant will then digest the unlucky fly.*

FOOD The kind of foods organisms eat varies considerably. You would probably not want to eat eucalyptus leaves, yet that is the only food a koala eats. A diet of wood may not seem tempting, but for the termite it is a source of energy and necessary chemical substances.

WATER Although you would probably not enjoy it, you could live for a week or more without food. But you would die in only a few days without water. It may surprise you to learn that 65 percent or more of your body is water. Other living things are also made up mainly of water.

In addition to making up much of your body, water serves many other purposes. Most substances dissolve in water. In this way, important chemicals can be transported easily throughout an organism. The blood of animals and the sap of trees, for example, are mainly water.

Most chemical reactions in living things cannot take place without water. Metabolism would come to a grinding halt without water. And it is water that carries away many of the metabolic waste products produced by living things. For green plants, water is also a raw material for photosynthesis (the food-making process).

Figure 1–23 *The gazelles and elephants drinking together at an African waterhole illustrate the importance of water to the survival of living things.*

Oxygen

You already know that for most living organisms oxygen is necessary for the process of respiration. Where do organisms get their oxygen? That depends on where they live. Organisms that live on land, whether plant or animal, obtain their oxygen directly from the air. Organisms that live under water either come up to the surface for oxygen (porpoises, for example) or remove the oxygen dissolved in the water (fish or seaweed, for example).

When organisms use oxygen they produce a waste product called carbon dioxide. For example, you breathe in oxygen when you inhale and breathe out carbon dioxide when you exhale. But the carbon dioxide is not wasted. Plants take carbon dioxide from the air and use it as a raw material in the process of photosynthesis. This cycling of oxygen and carbon dioxide is one reason living things have not used up all the available oxygen and carbon dioxide in the air.

Living Space

Do you enjoy the chirping of birds on a lovely spring morning? If you do, you may be surprised to learn that the birds are staking out their territory and warning intruders to stay away.

ACTIVITY
CALCULATING

You're All Wet

About 65 percent of your body mass is water.

Determine how many kilograms of water you contain.

Figure 1–24 *Organisms do not compete only for food and water. These elephants are crashing tusks in a power struggle for territory.*

Often there is a limited amount of food, water, and energy in an environment. As a result, only a limited number of the same kind of living thing can survive in a particular location. That is why many animals defend a certain area they consider to be their living space. The male sunfish, for example, defends its territory in ponds by flashing its colorful fins at other sunfish and darting toward any sunfish that comes too close. Coyotes howl at night to mark their territory and to keep other coyotes away. You might think of these behaviors as a kind of competition for living space.

Competition is the struggle among living things to get the proper amount of food, water, and energy. Animals are not the only competitors for these materials in their living space. Plants compete for sunlight and water as well. Smaller, weaker plants often die in the shadow of larger plants.

Proper Temperature

During the summer, temperatures as high as 58°C have been recorded on Earth. Winter temperatures can dip as low as –89°C. Most organisms cannot survive at such temperature extremes because many metabolic activities cannot occur at these temperatures. Without metabolism, an organism dies.

Actually, most organisms would quickly die at far less severe temperature extremes if it were not for **homeostasis** (hoh-mee-oh-STAY-sihs). Homeostasis is

Figure 1–25 *The coldblooded frilled-neck lizard is basking in the sun in order to achieve homeostasis. The warmblooded prairie dog relies on chemical activities in its body to maintain a constant body temperature.*

the ability of an organism to keep conditions inside its body the same, even though conditions in its external environment change. Maintaining a constant body temperature, no matter what the temperature of the surroundings, is part of homeostasis. Birds and certain other animals, such as dogs and horses, produce enough heat to keep themselves warm at low temperatures. When temperatures get too high, many birds make their throats flutter to cool off. Like dogs, some birds also pant to lower their body temperatures. Sweating has a similar effect on horses. Animals that maintain a constant body temperature are called warmblooded animals. Warmblooded animals can be active during both day and night, in hot weather and in cold.

Animals such as reptiles and fishes have body temperatures that can change somewhat with changes in the temperature of the environment. These animals are called coldblooded animals. But that does not mean that their blood is cold. Rather, it means that they must use their behavior to help maintain homeostasis. To keep warm, a coldblooded reptile, such as a crocodile, must spend part of each day lying in the sun. At night, when air temperature drops, so does the crocodile's body temperature. The crocodile becomes lazy and inactive. Coldblooded animals do not move around much at relatively high or low temperatures.

ACTIVITY
CALCULATING

Temperature Range

Calculate the range in temperature between the highest recorded temperature, 58°C and the lowest recorded temperature, −89°C.

What was the change in temperature in your area yesterday?

Find out the record high and low temperatures in your area. What is the difference in temperature?

1–3 Section Review

1. List and describe the basic needs of living things.
2. What is the relationship between the sun and the energy needed by living things?
3. Define homeostasis and explain its importance to living things.

Connection—*Ecology*
4. Thermal (heat) pollution occurs when heated water from factories and power plants is released into lakes and streams. Based on what you know about homeostasis, what problems to wildlife might result from thermal pollution?

CONNECTIONS

A Heated Experiment

The year was 1774. The English doctor Matthew Dobson decided to try a most unusual experiment. Even more unusual, he convinced four of his friends to help out. Dobson and his friends entered a heated room and sealed the door. While the temperature in the room rose higher and higher, the gleeful group continued to feed a fire blazing away in a stove in the room.

Five, ten, then twenty minutes passed. The room grew hotter and hotter. Dobson and his friends took their temperature every few minutes. To their surprise, even as raw eggs left in the room cooked before their eyes, their body temperatures barely rose at all. But they did sweat—oh, did they sweat!

After a while the adventurous group left the room. Once back in a more normal (and comfortable) condition, they began to analyze their observations and draw conclusions. Some of these conclusions may seem obvious to you, but to most people of that time they were significant discoveries. One conclusion was that sweating and the resulting *evaporation* of sweat was actually a way in which the body cooled itself. Of greater importance was the understanding that a healthy body keeps a stable internal temperature despite the outside environment. That is, the five men developed the idea that homeostasis is a basic requirement of life. From that concept came the notion that taking a person's temperature using a thermometer was one way of determining if that person was ill.

1–4 Chemistry of Living Things

What do foil wrap, a light-bulb filament, and a diamond have in common? These objects look very different and certainly have very different uses, but all are examples of **elements.** An element is a pure substance that cannot be broken down into any simpler substances by ordinary means.

When two or more elements are chemically joined together, **compounds** are formed. Water is a compound made of the elements hydrogen and oxygen. Table salt, which you may use to flavor your food, is a compound made of sodium and chlorine. Sand and glass are compounds composed of the elements silicon and oxygen. There are thousands of different compounds all around you. In fact, you are made of many compounds. Scientists classify compounds into two groups.

Inorganic Compounds

Compounds that may or may not contain the element carbon are called inorganic compounds. Most inorganic compounds do not contain carbon. However, carbon dioxide is an exception. Table salt, ammonia, rust, and water are inorganic compounds.

Organic Compounds

Most of the compounds in living things contain carbon, which is usually combined with other elements such as hydrogen and oxygen. These compounds are called **organic compounds.** The term organic refers to life. Because some of these compounds were first discovered in living things, they were appropriately named organic compounds.

Figure 1–26 *Nonliving things, such as these amethyst crystals, are made of inorganic compounds. Living things, such as the mushroom and green plants, are made primarily of organic compounds. What element is found in all organic compounds?*

Figure 1–27 *A thick layer of fat and a nice fur coat keep this Black bear warm throughout the year.*

There are more than 3 million different organic compounds in living things. **Organic compounds that are basic to life include carbohydrates, fats and oils, proteins, enzymes, and nucleic acids.**

CARBOHYDRATES The main source of energy for living things is **carbohydrates.** Carbohydrates are made of the elements carbon, hydrogen, and oxygen. Sugar and starch are two important carbohydrates. Many fruits are high in sugar content. Potatoes, rice, noodles, and bread are common sources of starch. What are some foods you eat that contain sugars and starches?

Carbohydrates are broken down inside the body into a simple sugar called glucose. The body then uses glucose to produce the energy needed for life activities. If an organism has more sugar than it needs for its energy requirements, it will store the sugar for later use. The sugar is stored as starch. Starch, then, is a stored form of energy.

FATS AND OILS Another group of energy-rich compounds made of carbon, hydrogen, and oxygen are **fats** and **oils.** The more proper scientific term for these compounds is lipids. How can you tell a fat from an oil? Actually, it is quite easy. Fats are solid at room temperature; oils are liquid at room temperature.

PROTEINS Like carbohydrates and fats, **proteins** are organic compounds made up of carbon, hydrogen, and oxygen. But proteins also contain the element nitrogen and sometimes the elements sulfur and phosphorus. Some important sources of proteins are eggs, meat, fish, beans, nuts, and poultry.

The building blocks of proteins are **amino acids.** There are about 20 different amino acids. But because amino acids combine in many ways, they form thousands of different proteins.

Proteins perform many jobs for an organism. They are necessary for the growth and repair of body structures. Proteins are used to build body parts such as hair and muscles. Proteins provide energy. Some proteins, such as those in blood, carry oxygen throughout the body. Other proteins fight germs that invade the body. Still other proteins make chemical substances (hormones) that start, stop, and regulate many important body activities.

ENZYMES   A special type of protein that regulates chemical activities within the body is called an **enzyme.** Enzymes act as catalysts. A catalyst is a substance that speeds up or slows down chemical reactions but is not itself changed by the reaction. Without enzymes, the chemical reactions of metabolism could not take place or would occur so slowly that they would be of little help to the organism.

NUCLEIC ACIDS   Do you remember the "blueprints" of life we discussed earlier in this chapter? These blueprints are organic chemicals called **nucleic acids.** Nucleic acids, which are very large compounds, store information that helps the body make the proteins it needs. The nucleic acids control the way the amino acids are put together so that the correct protein is formed. This process is similar to the way a carpenter uses a blueprint to build a house. Can you now understand why nucleic acids are called the blueprints of life?

One nucleic acid is **DNA,** or deoxyribonucleic (dee-ahks-ih-right-boh-noo-KLEE-ihk) acid. DNA stores the information needed to build a protein. DNA also carries "messages" about an organism that are passed from parent to offspring.

Another nucleic acid is called **RNA,** or ribonucleic (right-boh-noo-KLEE-ihk) acid. RNA "reads" the message carried by DNA and guides the protein-making process. Together, these two nucleic acids contain the information and carry out the steps that make each organism what it is.

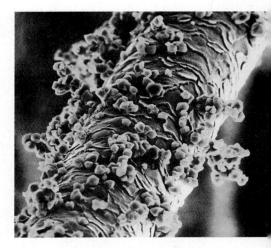

Figure 1–28 *This shaft of hair is composed primarily of proteins. The green droplets are the result of spraying the hair with dry shampoo.*

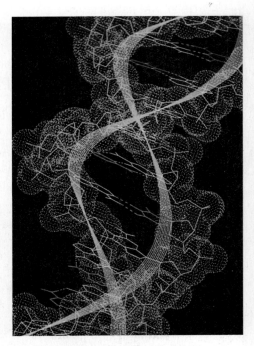

Figure 1–29 *Nucleic acids have been called the blueprints of life. The computer-generated image shows the structure of a DNA molecule, the molecule that controls protein production and heredity.*

1–4 Section Review

1. Compare an element and a compound.
2. List and describe the basic organic compounds necessary for life.
3. Why is it important that enzymes are not changed by the reactions they control?

Connection—*Nutrition*
4. Based on your knowledge of the chemistry of living things, explain why it is important to eat a balanced diet.

Laboratory Investigation

You Are What You Eat

Problem

Does your school lunch menu provide a balanced diet?

Materials (per group)

> school lunch menu for the current week
> pencil
> paper
> reference book or textbook on nutrition

Procedure

1. Obtain a copy of your school's lunch menu for one week.

2. Make a table similar to the one shown here that lists the four basic food groups: meat group, vegetable-fruit group, milk group, bread-cereal group.

3. For each day of the week, place each item from the menu in the appropriate food group in your table. An example has been provided in the sample table.

4. Make a second table similar to the one shown here that lists the three major nutrients: carbohydrates, fats, and proteins. List those foods containing large amounts of these nutrients under the proper heading in your table. An example has been provided.

5. On a third table similar to the one shown here, identify those foods that are plants or plant products and those that are animals or animal products. An example has been provided.

Observations

Study the data you have collected and organized.

Analysis and Conclusions

1. What conclusions can you draw regarding your school's lunch program?

2. According to your data, do the foods represent a balanced diet? Do foods in certain categories appear much more often than foods in other categories?

3. What changes, if any, would you make in the menus?

4. **On Your Own** Do a similar exercise, but this time analyze the dinners you eat for a week.

Meat Group	Vegetable-Fruit Group	Milk Group	Bread-Cereal Group
Hamburger			Roll

Carbohydrates	Fats	Proteins
Roll	Hamburger	Hamburger

Plants or Plant Products	Animals or Animal Products
Roll	Hamburger

Summarizing Key Concepts

1–1 The Origin of Life

▲ Scientists have shown that some of the substances that make up living things could have formed on Earth some 4 billion years ago.

▲ The first living things to evolve on Earth were single-celled organisms.

▲ Early cells were consumers, feeding off the soup of chemicals in which they floated. In time, cells that could perform photosynthesis evolved.

▲ Over time, early cells developed the ability to use oxygen in their metabolic pathways.

1–2 Characteristics of Living Things

▲ Living things are cellular; that is, they are made up of one or more cells.

▲ Metabolism is the sum of all chemical activities essential to life.

▲ A living thing reacts to a stimulus, which is a change in the environment, by producing a response.

▲ Reproduction is the process by which organisms produce offspring. Reproduction may be asexual or sexual, depending on the organism.

1–3 Needs of Living Things

▲ Living things need energy for metabolism. The primary source of energy for almost all living things is the sun.

▲ All living things need food and water.

▲ Oxygen in the air or dissolved in water is used by organisms during respiration.

▲ All living things need living space that provides adequate food, water, and energy.

▲ Homeostasis is the ability of an organism to keep conditions constant inside its body when the outside environment changes.

1–4 Chemistry of Living Things

▲ Most inorganic compounds do not contain the element carbon. Organic compounds do contain carbon. The organic compounds important to life are carbohydrates, fats and oils, proteins, enzymes, and nucleic acids.

▲ DNA and RNA are the nucleic acids that carry information that controls the building of proteins. DNA also is considered the "blueprint" for life as it directs the development of an organism's offspring.

Reviewing Key Terms

Define each term in a complete sentence.

1–2 Characteristics of Living Things

spontaneous
  generation
metabolism
ingestion
digestion
respiration
excretion
life span
stimulus

response
sexual reproduction
asexual reproduction

1–3 Needs of Living Things

homeostasis

1–4 Chemistry of Living Things

element

compound
organic compound
carbohydrate
fat
oil
protein
amino acid
enzyme
nucleic acid
DNA
RNA

Chapter Review

Content Review

Multiple Choice

Choose the letter of the answer that best completes each statement.

1. Earth's early atmosphere was changed dramatically by the evolution of
   a. amino acids.
   b. sexual reproduction.
   c. photosynthesis.
   d. asexual reproduction.
2. The theory that life could spring from nonliving matter is called
   a. spontaneous generation.
   b. asexual reproduction.
   c. homeostasis.
   d. stimulus/response.
3. The building up and breaking down of chemical substances necessary for life is called
   a. respiration.   c. digestion.
   b. ingestion.   d. metabolism.
4. The complex compound that carries the information needed to make proteins is
   a. carbohydrate.   c. DNA.
   b. enzyme.   d. lipid.

5. The struggle among living things to obtain the resources needed for survival is called
   a. spontaneous generation.
   b. competition.
   c. homeostasis.
   d. metabolism.
6. Which of these is a substance made of a single element?
   a. table salt   c. diamond
   b. water   d. glass
7. Carbohydrate is to glucose as
   a. fat is to oil.
   b. protein is to amino acid.
   c. DNA is to RNA.
   d. hydrogen is to oxygen.
8. The process of combining oxygen with the products of digestion to produce energy is called
   a. respiration.   c. homeostasis.
   b. excretion.   d. reproduction.

True or False

If the statement is true, write "true." If it is false, change the underlined word or words to make the statement true.

1. Planet Earth is about <u>4.6 million years</u> old.
2. The theory of <u>spontaneous generation</u> states that life can spring from nonliving matter.
3. The process by which an organism puts food into its body is <u>digestion</u>.
4. Green plants produce the waste product <u>carbon dioxide</u> during photosynthesis.
5. The process of getting rid of body wastes is called <u>digestion</u>.
6. Ammonia and water are examples of <u>inorganic compounds</u>.
7. Carbohydrates include <u>sugars</u> and <u>starches</u>.

Concept Mapping

Complete the following concept map for Section 1–2. Refer to pages D6–D7 to construct a concept map for the entire chapter.

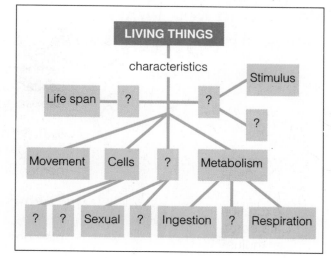

Concept Mastery

Discuss each of the following in a brief paragraph.

1. Describe at least five ways in which you can tell a living thing from a nonliving thing.
2. Your friend insists that plants are not alive because they do not move. Give specific examples to explain why your friend is wrong.
3. Describe four examples of metabolic processes.
4. Defend this statement: All plants and animals get their energy from the sun.

Critical Thinking and Problem Solving

Use the skills you have developed in this chapter to answer each of the following.

1. **Drawing conclusions** Is a peach pit living or nonliving? Explain your answer.
2. **Relating concepts** Why is the study of chemistry important to the understanding of living things?
3. **Relating cause and effect** In the days when people believed in spontaneous generation, one scientist developed the following recipe for producing mice: Place a dirty shirt and a few wheat grains in an open pot; wait three weeks. Suggest a reason why this recipe may have worked. How could you prove that spontaneous generation was not responsible for the appearance of mice?
4. **Applying concepts** Figure 1–10 on page 20 shows three sets of jars that illustrate Redi's experiment. Explain why the second set of jars did not provide enough evidence to disprove the spontaneous-generation theory.
5. **Making comparisons** Compare the growth of a sand dune to the growth of a living organism.
6. **Synthesizing data** Which is more likely to result in increased variety among organisms, sexual reproduction or asexual reproduction?
7. **Designing an experiment** A plant salesperson tells you that plants respond to the stimulus of classical music by growing more quickly. Design an experiment to test the salesperson's claim.
8. **Using the writing process** Develop a poster or an advertising campaign that explains the importance of eating a balanced diet in terms of the needs of living things.

Cell Structure and Function

When it first appeared on Earth about 3.5 billion years ago, it was a tiny structure made up of tinier parts alive with activity. Over the course of millions of years, it changed a bit here and a bit there. As new parts evolved, it became able to do increasingly different jobs: It could build complex chemicals, it could release bursts of energy, and eventually it could even move by itself.

Today these tiny structures still exist. They are a bridge to the distant past. They are the building blocks of all living things. They are cells!

Cells are fascinating and, in many ways, mysterious objects. Scientists probe the secrets of cells much as explorers journeying through parts of an uncharted world do. Read on and you too will become an explorer as you take a fantastic journey through the microscopic world of the cell.

Journal *Activity*

You and Your World Living things are made of cells. Cells are the building blocks of living things. This textbook is made of atoms. Atoms are the building blocks of matter. In your journal, describe in words or pictures the characteristics of cells that distinguish them from atoms. That is, why are cells alive and atoms are not? When you complete this chapter, go back to your journal and make any changes you feel are necessary.

◀ *The round objects in the photograph are white blood cells located in the thymus, an organ in which certain white blood cells are produced. A computer has highlighted the cells, giving them their green appearance.*

2-1 The Cell Theory

The basic units of structure and function of living things are **cells.** Most cells are too small to be seen with the unaided eye. As a result, many of the even smaller structures that make up a cell remained a mystery to scientists for hundreds of years. The structures that make up a cell are called **organelles,** which means tiny organs. The organelles were not revealed until the seventeenth century, when the first microscopes were invented.

In 1665, while looking at a thin slice of cork through a compound microscope, the English scientist Robert Hooke observed tiny roomlike structures. He called these structures cells. But the cells that Hooke saw in the slice of cork were not alive. What Hooke saw were actually the outer walls of dead plant cells.

At about the same time, Anton van Leeuwenhoek (LAY-vuhn-hook), a Dutch fabric merchant and amateur scientist, used a simple microscope to examine materials such as blood, rainwater, and scrapings from his teeth. In each material, van Leeuwenhoek observed living cells. He even found tiny living things in a drop of rainwater. Van Leeuwenhoek called these living things "animalcules." The smallest of the animalcules observed by van Leeuwenhoek are today known as bacteria. Bacteria are single-celled organisms. These discoveries made van Leeuwenhoek famous all over the world.

During the next two hundred years, new and better microscopes were developed. Such microscopes made it possible for the German botanist Matthias Schleiden to view different types of plant parts. Schleiden discovered that all the plant parts he examined were made of cells. One year later, the German zoologist Theodor Schwann made similar observations using animal parts. About twenty years later, a German physician named Rudolph Virchow discovered that all living cells come only from other living cells.

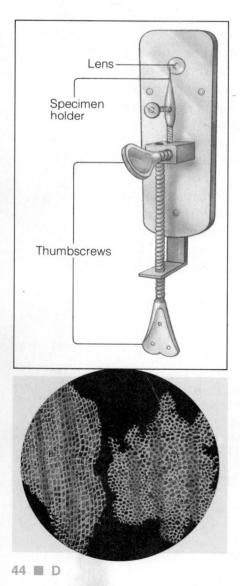

Lens

Specimen
holder

Thumbscrews

Figure 2–1 *Van Leeuwenhoek's simple microscope (top) could magnify objects a few hundred times. Robert Hooke made this drawing of cork cells (bottom) using a microscope he built. Hooke was not looking at living cells, but rather at the cell walls that surround living cork cells.*

The work of Schleiden, Schwann, Virchow, and other biologists led to the development of the **cell theory,** which is one of the cornerstones of modern biology. **The cell theory states that**

- **All living things are made of cells.**
- **Cells are the basic units of structure and function in living things.**
- **Living cells come only from other living cells.**

2–1 Section Review

1. What is the cell theory?
2. What term is used for the structures that make up a cell?

Connection—*Science and Technology*
3. Discuss the relationship between technology and the development of the cell theory.

2–2 Structure and Function of Cells

You are about to take an imaginary journey. It will be quite an unusual trip because you will be traveling inside a living organism, visiting its tiny cells. On your trip you will be observing some of the typical structures found in plant and animal cells.

All living things are made of one or more cells. As you have just learned, cells are the basic units of structure and function in living things. Most cells are much too small to be seen without the aid of a

Guide for Reading

Focus on these questions as you read.

▶ *What structures are found within a typical cell and what function do they serve?*

▶ *What are the five levels of organization in multicellular organisms?*

Plant Cells

1. To view a plant cell, remove a very thin transparent piece of tissue from an onion.

2. Place the onion tissue on a glass slide.

3. Add a drop of iodine stain to the tissue and cover with a coverslip.

4. Observe the onion tissue under low power and under high power of your microscope.

Draw a diagram of an onion cell and label its parts.

Figure 2–3 *The cell wall gives support and protection to plant cells, enabling giant redwoods to grow tall and straight.*

microscope. In fact, most cells are smaller than the period at the end of this sentence. (One exception is the yolk of an egg, which is actually a large single cell.) Within a cell are even smaller structures called organelles. **The structures within a cell function in providing protection and support, forming a barrier between the cell and its environment, building and repairing cell parts, transporting materials, storing and releasing energy, getting rid of waste materials, and increasing in number.**

Whether found in an animal or in a plant, most cells share certain similar characteristics. It is these characteristics that you are going to learn about. So hop aboard your imaginary ship and prepare to enter a typical plant cell. You will begin by sailing up through the trunk of an oak tree. Your destination is that box-shaped structure directly ahead. See Figure 2–4.

Cell Wall: Support and Protection

Entering the cell of an oak tree is a bit difficult. First you must pass through the **cell wall.** Strong and stiff, the cell wall is made of cellulose, a nonliving material. Cellulose is a long chain of sugar molecules that the cell manufactures. (The stringy part of celery is cellulose found in the cell walls of the celery stalk.)

The rigid cell wall is found in plant cells, but not in animal cells. The cell wall helps to protect and support the plant so that it can grow tall. Think for a moment of grasses, trees, and flowers that support themselves upright. No doubt you can appreciate the important role the cell wall plays for the individual cell and for the entire plant.

Although the cell wall is stiff, it does allow water, oxygen, carbon dioxide, and certain dissolved materials to pass into and out of the cell. So sail on through the cell wall and enter the cell.

Cell Membrane: Doorway of the Cell

As you pass through the cell wall, the first structure you encounter is the **cell membrane.** In a plant cell, the cell membrane is just inside the cell wall. In an animal cell (which has no cell wall), the cell membrane forms the outer covering of the cell.

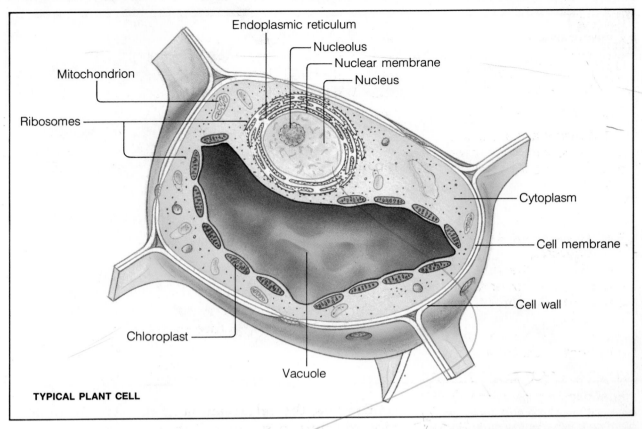

Endoplasmic reticulum
Nucleolus
Nuclear membrane
Nucleus

Mitochondrion

Ribosomes

Cytoplasm

Cell membrane

Cell wall

Chloroplast

Vacuole

TYPICAL PLANT CELL

Figure 2–4 *A typical plant cell contains many different structures, each having a characteristic shape and function. What is the outer barrier surrounding a plant cell called?*

The cell membrane has several important jobs. One of these important jobs is to provide protection and support for the cell. Unlike a plant cell, an animal cell does not have a rigid cell wall. Instead, an animal's cell membrane contains a substance called cholesterol that strengthens the cell membrane. As you might expect, a plant's cell membrane does not contain cholesterol.

As your ship nears the edge of the cell membrane, you notice that there are tiny openings, or pores, in the membrane. You steer toward an opening. Suddenly your ship narrowly misses being struck by a chunk of floating waste material passing out of the cell. You have discovered another job of the cell membrane. This membrane helps to control the movement of materials into and out of the cells. You will learn more about how materials pass through the cell membrane in Chapter 3.

In a sense, the cell membrane is like the walls that surround your house or your apartment. Just as the walls of your home form a barrier between you and the outside world, the cell membrane forms a barrier between the living material inside the cell.

Activity Bank

Now You See It—Now You Don't, p. 107

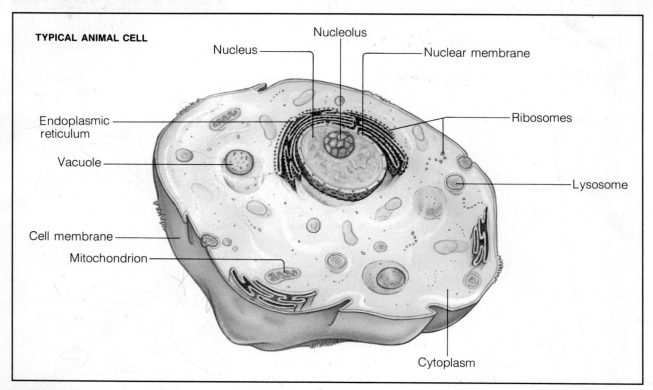

TYPICAL ANIMAL CELL

Nucleus

Nucleolus

Nuclear membrane

Endoplasmic reticulum

Ribosomes

Vacuole

Lysosome

Cell membrane

Mitochondrion

Cytoplasm

Figure 2–5 *An animal cell has many of the same structures as a plant cell. What is the outer barrier surrounding an animal cell called?*

and the outside environment of the cell. No doubt you would be quite unhappy if nothing could get into and out of your home. After all, you need to have food, water, and electricity come into your home. And you need to have waste products removed from your home before they build up. To ensure the survival of the cell, the cell membrane must allow materials to pass into and out of the cell. So you can think of the cell membrane as a barrier with doorways.

Everything the cell needs, from food to oxygen, enters the cell through the cell membrane. And harmful waste products exit through the cell membrane as well. In this way, the cell stays in smooth-running order, keeping conditions inside the cell the same even though conditions outside the cell may change. As you may recall from Chapter 1, the ability to maintain a stable internal environment, or homeostasis, is one of the important needs of all living things. Now sail on through a doorway in the cell membrane and enter a living cell.

Nucleus: Control Center of the Cell

As you sail inside the cell, a large, oval structure comes into view. This structure is the control center of the cell, or the **nucleus** (NOO-klee-uhs). The nucleus acts as the "brain" of the cell, regulating or controlling all the activities of the cell. See Figure 2–7.

NUCLEAR MEMBRANE Like the cell itself, the nucleus is also surrounded by a membrane. As you might expect, it is called the nuclear membrane. This membrane is similar to the cell membrane in that it allows materials to pass into or out of the nucleus. Small openings, or pores, are spaced regularly around the nuclear membrane. Each pore acts as a passageway. So set your sights for that pore just ahead and carefully glide into the nucleus.

CHROMOSOMES Those thick, rodlike objects floating directly ahead in the nucleus are **chromosomes.** Steer carefully to avoid colliding with the delicate chromosomes. For it is the chromosomes that direct all the activities of the cell, including growth and reproduction. In addition, chromosomes are responsible for passing on the traits of the cell to new cells. Chromosomes, for example, make sure that skin cells grow and divide into more skin cells.

The large, complex molecules that make up the chromosomes are compounds called nucleic acids. Nucleic acids store the information that helps a cell make the proteins it requires. And proteins are necessary for life. Some proteins are used to form parts of the cell, such as the cell membrane. Other

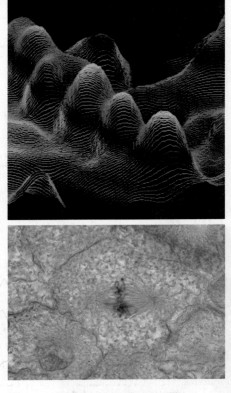

Figure 2–6 *Notice the rodlike chromosomes in the nucleus (bottom). This unusual photograph is an image of DNA in a chromosome, produced by a scanning electron microscope (top). The colors have been added by a computer.*

Figure 2–7 *The nucleus directs all the activities of a cell. Notice the various structures that make up the nucleus and the appearance of the nucleus in a liver cell viewed through an electron microscope.*

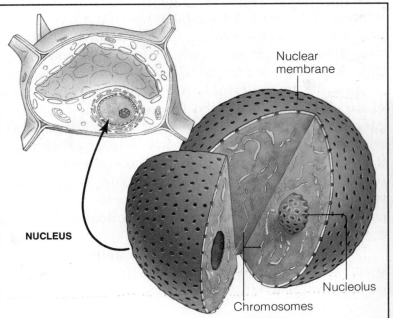

Nuclear membrane

NUCLEUS

Nucleolus

Chromosomes

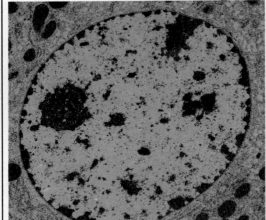

proteins make up different enzymes and hormones used inside and outside the cell. Enzymes and hormones regulate cell activities.

The two nucleic acids found in cells are DNA and RNA. In Chapter 1 we called the nucleic acids the carriers of the blueprints of life. Working together, DNA and RNA store the information and carry out the steps in the protein-making process necessary for life. The DNA remains in the nucleus. But the RNA, carrying its protein-building instructions, leaves the nucleus through pores in the nuclear membrane. So hitch a ride on the RNA leaving the nucleus and continue your exploration of the cell.

NUCLEOLUS As you prepare to leave the nucleus, you spot a small object floating past. It is the nucleolus (noo-KLEE-uh-luhs), or "little nucleus." For many years the function of the nucleolus remained something of a mystery to scientists. Today it is believed that the nucleolus is the site of ribosome production. Ribosomes, as you will soon learn, are involved in the protein-making process in the cell.

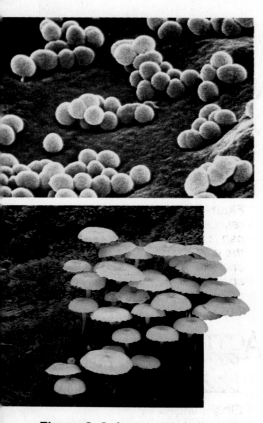

Figure 2–8 *As you read about the structures in a typical cell, keep in mind that organisms do differ in cell structure. Bacteria, members of the kingdom Monera, do not contain a distinct nucleus (top). Fungi, which look similar to plants, have nuclei but do not always have cells separated by a cell wall. For this reason, fungi are placed in the kingdom Fungi (bottom).*

Endoplasmic Reticulum: Transportation System of the Cell

As you leave the nucleus, you find yourself floating in a clear, thick, jellylike substance called the **cytoplasm.** The cytoplasm is the term given to the region between the nucleus and the cell membrane. While you are in the cytoplasm your ship needs no propulsion. For the cytoplasm is constantly moving, streaming throughout the cell. Many of the important cell organelles are located within the cytoplasm.

Steering in the cytoplasm is a bit difficult because of the organelles scattered throughout. The first organelle you encounter as you sail out of the nucleus is a maze of tubular passageways. These passageways lead out from the nuclear membrane. Some of the passageways lead to the cell membrane. Others lead to all the other areas of the cell. These clear, tubular passageways form the **endoplasmic reticulum** (en-doh-PLAZ-mihk rih-TIHK-yuh-luhm).

The endoplasmic reticulum is a transportation system. Its network of passageways spreads throughout the cell, carrying proteins from one part of the

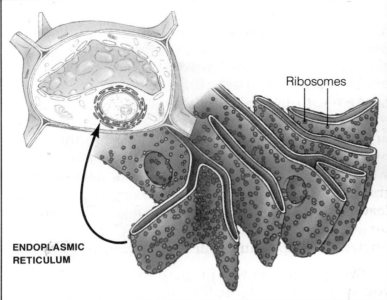

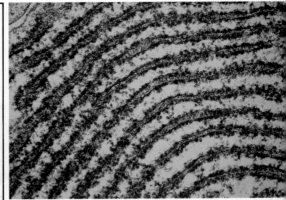

Figure 2–9 *The endoplasmic reticulum is a canal system that can transport proteins throughout the cell. In this photograph of the endoplasmic reticulum, the dark spots are ribosomes, the sites of protein production.*

cell to another—or from the cell through the cell membrane and out the cell. If you look at Figure 2–9, you will see that the endoplasmic reticulum is well suited for its transportation job.

Ribosomes: Protein Factories of the Cell

Steer your ship directly into the endoplasmic reticulum passageways. From here you can travel anywhere you want in the cell. Before moving on, however, look closely at the inner surface of the endoplasmic passageways. Attached to the surface are grainlike bodies called **ribosomes.** Recall that ribosomes are produced in the nucleolus. From the nucleolus they pass out of the nucleus. Many of them end up attached to the inner lining of the endoplasmic reticulum.

Ribosomes, which are made primarily of the nucleic acid RNA, are the protein-making sites of the cell. The RNA in the ribosomes, along with the RNA sent out from the nucleus, directs the production of proteins. (Keep in mind that the production of RNA is controlled by the DNA in the chromosomes. So the DNA in the chromosomes is the real control center of the cell.)

It is no surprise that many ribosomes are found in the endoplasmic reticulum. This is a perfect location for them. For, once the ribosomes have made

the proteins needed by the cell, they can immediately drop them off into the passageways of the endoplasmic reticulum. From there the proteins can be transported to any part of the cell where they are needed—or out of the cell if necessary.

As you leave the endoplasmic reticulum, you notice that not all ribosomes are attached to the endoplasmic reticulum. Some float freely in the cytoplasm. Watch out! There go a few passing by. The cell you are in seems to have many ribosomes. What might this tell you about its protein-making activity?

Mitochondria: Powerhouses of the Cell

As you pass by the ribosomes, you see other structures looming ahead. These structures are called **mitochondria** (might-oh-KAHN-dree-uh; singular: mitochondrion). Mitochondria supply most of the energy for the cell. Somewhat larger than the ribosomes, these rod-shaped structures are often referred to as the "powerhouses" of the cell. See Figure 2–10.

Inside the mitochondria, simple food substances such as sugars are broken down into water and carbon dioxide gas. Large amounts of energy are released during the breakdown of sugars. The mitochondria gather this energy and store it in special energy-rich molecules. These molecules are convenient energy packages that the cell uses to do

Figure 2–10 Mitochondria are the powerhouses of the cell. They provide the cell with the energy it needs to survive. Note the structure of the mitochondrion in the diagram and in the electron micrograph.

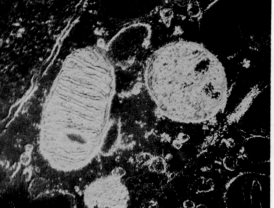

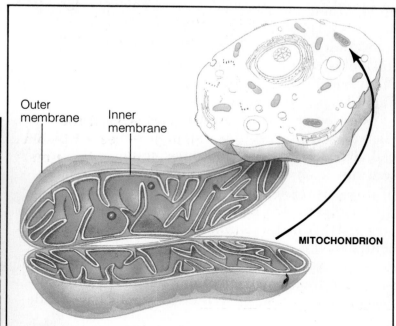

Outer membrane

Inner membrane

MITOCHONDRION

all its work. The more active the cell, the more mitochondria it has. Some cells, such as human liver cells, contain more than 1000 mitochondria. You will read more about mitochondria and energy in Chapter 4.

Because mitochondria have a small amount of their own DNA, scientists hypothesize that mitochondria were once tiny living organisms. These organisms, it is believed, invaded other cells millions of years ago. The DNA molecules in the mitochondria were passed from one generation of cells to the next as less complex organisms evolved into more complex organisms. (Keep in mind that even the smallest cells are still quite complex.) Now all living cells contain mitochondria. No longer invaders, mitochondria are an important part of living cells.

Vacuoles: Storage Tanks for Cells

Steer past the mitochondria and head for that large, round, water-filled sac floating in the cytoplasm. This sac is called a **vacuole** (VA-kyoo-ohl). Most plant cells and some animal cells have vacuoles. Plant cells often have one very large vacuole. Animal cells, if they contain any vacuoles, generally have a few small ones.

Vacuoles act like storage tanks. Food and other materials needed by the cell are stored inside the vacuoles. Vacuoles can also store waste products. In plant cells, vacuoles are the main water-storage areas. When water vacuoles in plant cells are full, they swell and make the cell plump. This plumpness keeps a plant firm.

Lysosomes: Cleanup Crews for the Cell

If you carefully swing your ship around the vacuole, you may be lucky enough to see a lysosome (LIGH-suh-sohm). Lysosomes are common in animal cells but are not often observed in plant cells.

Lysosomes are small, round structures involved with the digestive activities of the cell. See Figure 2–13 on page 54. Lysosomes contain enzymes that break down large food molecules into smaller ones. These smaller food molecules are then passed on to the mitochondria, where they are "burned" to provide energy for the cell.

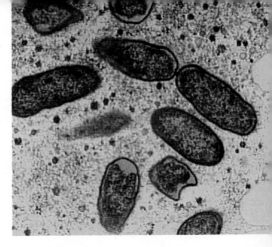

Figure 2–11 *Before they became permanent members of living cells, ancient mitochondria may have been similar in structure to these rickettsia. Mitochondria are no longer outside invaders. The rickettsia shown here, however, cause a serious disease called Rocky Mountain Spotted Fever.*

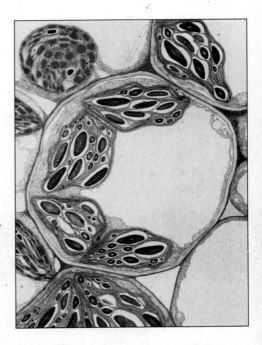

Figure 2–12 *The large, roundish, empty spaces in these plant cells are vacuoles. What materials do vacuoles store?*

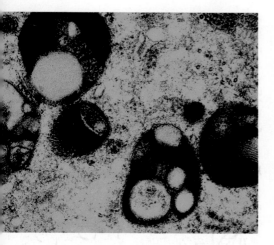

Figure 2–13 *These spherical organelles are lysosomes that have been magnified approximately 95,000 times. Lysosomes contain enzymes that can digest other organelles that have outlived their usefulness.*

Lysosomes are not involved just in digesting food. Many parts of the cell age and outlive their usefulness. One task of lysosomes is to digest old cell parts, releasing the substances in those aging cell parts so that they can be used again to build new parts. In this sense, you can think of lysosomes as the cell's cleanup crew.

Although lysosomes contain powerful digestive enzymes, you need not worry about your ship's safety. The membrane surrounding a lysosome keeps the enzymes from escaping and digesting the entire cell! Lysosomes can, however, digest whole cells when the cells are injured or dead. In an interesting process in the growth and development of a tadpole into a frog, lysosomes in the tadpole's tail cells digest the tail. Then the material is reused to make new body parts for the frog.

Chloroplasts: Energy Producers for the Cell

Your journey through the cell is just about over. But before you leave, look around you once again. Have you noticed any large, irregularly shaped green structures floating in the cytoplasm? If so, you have observed **chloroplasts.** Chloroplasts are green because they contain a green pigment called chlorophyll. Chlorophyll captures the energy of sunlight, which can then be used to help produce food for

Figure 2–14 *Chloroplasts are organelles that use sunlight to produce food in a process called photosynthesis. Would you be likely to find a chloroplast in an animal cell?*

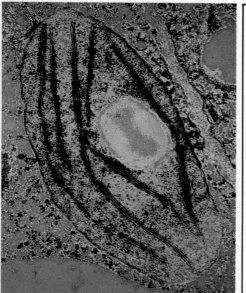

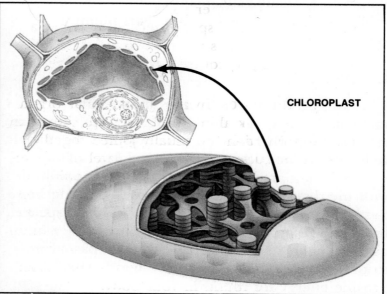

CHLOROPLAST

the plant cell. This process is called photosynthesis. You will read more about chloroplasts and photosynthesis in Chapter 4.

Cell Specialization

You have just read about the structures found in plant and animal cells. These structures help to keep the cell alive and functioning properly. In unicellular organisms such as bacteria, the single cell performs all the functions necessary for life. But, as you know, many organisms (including yourself) are multicellular. In multicellular organisms, each cell may well perform a specialized function for the entire organism. That is, the cell not only completes all its own life activities, it also contributes to the life of the organism. Without cell specialization, the evolution of multicellular organisms could not have occurred.

Tissues, Organs, and Organ Systems

You have just read that cells in multicellular organisms are specialized to perform specific tasks for the organism. So it should not surprise you to learn that cells are often organized in order to better serve the needs of the organism. In other words, within a multicellular organism there is a **division of labor.** Division of labor means that the work of keeping the organism alive is divided among the different parts of the body. Each part has a specific job to do. And as each part does its special job, it works in harmony with all the other parts.

The arrangement of specialized parts within a living thing is sometimes referred to as levels of organization. Cells, of course, are the first level of organization.

TISSUES: LEVEL TWO In any multicellular organism, cells rarely work alone. Cells that are similar in structure and function are usually joined together to form **tissues.** Tissues are the second level of organization. What is the first level of organization?

For example, bone cells in your body form bone tissue, a strong, solid tissue that gives you shape and support. Blood cells in your body are part of blood tissue, a liquid tissue responsible for transporting food and oxygen throughout the body. What other types of tissues are found in your body?

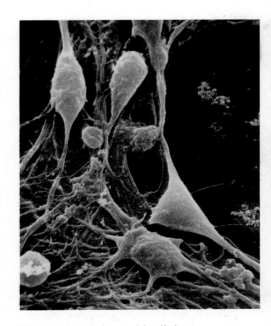

Figure 2–15 *In multicellular organisms, many cells are specialized to perform a specific function for the organism. Here you see nerve cells (neurons) in the part of the brain used for thinking skills such as reading and understanding this textbook.*

ACTIVITY

WRITING

Word Clues

The definition of a word can often be determined by knowing the meaning of its prefix. For the following words, look up the prefix in each and tell how it relates to the definition:

chloroplast
mitochondria
chromosome
lysosome
cytoplasm

ORGAN SYSTEMS

System	Function
Skeletal	Protects and supports the body
Muscular	Supports the body and enables it to move
Skin	Protects the body
Digestive	Receives, transports, breaks down, and absorbs food throughout the body
Circulatory	Transports oxygen, wastes, and digested food throughout the body
Respiratory	Permits the exchange of gases in the body
Excretory	Removes liquid and solid wastes from the body
Endocrine	Regulates various body functions
Nervous	Conducts messages throughout the body to aid in coordination of body functions
Reproductive	Produces male and female sex cells

Figure 2–16 *Each organ system is made of a group of organs that work together. Which organ system enables the body to move?*

ORGANS: LEVEL THREE   In general, tissues are further organized into **organs,** the third level of organization in living things. Organs are groups of different tissues that work together. Your heart, for example, is an organ made up of muscle tissue, blood tissue, and nerve tissue. You are probably familiar with the names of many of the body organs. The brain, stomach, kidneys, and skin are some examples. Can you name others?

ORGAN SYSTEMS: LEVEL FOUR   Like cells and tissues, organs seldom work alone. They "cooperate" with one another and form specific **organ systems.** Organ systems are the fourth level of organization in living things. An organ system is a group of organs working together to perform a specific function for the organism.

Figure 2–16 shows the various organ systems found in many animals, including yourself. Study Figure 2–16 until you understand the function of each organ system in your body.

ORGANISMS: LEVEL FIVE   You are an organism. Dogs, trees, and buttercups are also organisms. Even a unicellular bacterium is an organism. An organism is an entire living thing that carries out all the basic life functions. The organism is the fifth and highest level of organization.

Cells, tissues, organs, organ systems, organisms— by now one thing should be clear to you: Each level of organization interacts with every other level. And the smooth functioning of a complex organism is the result of all its various parts working together.

2–2 Section Review

1. List and describe the organelles found in a typical plant cell. Make note of any organelles that are not found in an animal cell as well.
2. List and describe the five levels of organization in living things.

Connection—*You and Your World*
3. Compare each of the cell organelles to a structure in your house or apartment. Give an explanation for each comparison.

Cellular Communication

Smog! Acid rain! You hear about these *pollution* problems almost everyday. One of the substances that leads to both smog and acid rain is nitric oxide (a molecule containing nitrogen and oxygen). Nitric oxide is released in large quantities from car exhaust pipes.

Because of the pollution problems associated with it, nitric oxide is high on environmentalists' list of substances that need to be eliminated from our atmosphere. Strangely enough, however, in 1991 scientists discovered that nitric oxide is one of the most important chemicals in the human body! One of its most significant functions involves communication between cells.

It may surprise you to know that cells can communicate with one another, particularly with nearby cells. As a result of this communication, the actions of cells are well coordinated in the body. It turns out that a good deal of cellular communication is controlled by nitric oxide. In the brain, for example, messages transmitted between nerve cells allow you to think, coordinate movement, feel emotions, and so on. It now appears as if the nerve cells of

the brain have small amounts of nitric oxide that help to transmit messages. Scientists also suspect that too much nitric oxide in the brain can lead to disorders such as Huntington's disease and possibly Alzheimer's disease.

Blood vessels also communicate using nitric oxide. When cells in the walls of blood vessels release nitric oxide, other cells relax. This relaxation helps the body lower blood pressure. Scientists now predict that medicines using nitric oxide might soon be used to control blood pressure.

The list of cells that may communicate using nitric oxide is long. As Dr. Salvador Moncada, a scientist in London, states, "Nitric oxide may be a universal signal transducer." What that means is that nitric oxide may be used throughout the body to send messages from one cell to another. Why, you may wonder, did it take so long for scientists to discover such a common substance in the body? The answer is actually quite simple. In the body, nitric oxide lasts for about five seconds before it is broken down. A rather important five seconds, indeed!

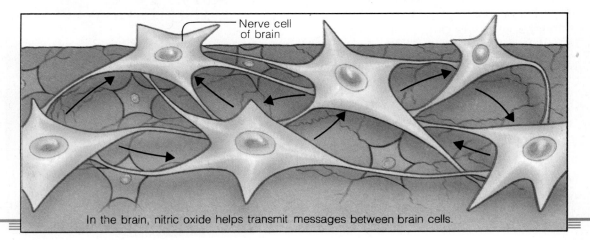

In the brain, nitric oxide helps transmit messages between brain cells.

Laboratory Investigation

Using the Microscope

Problem

How do you use the microscope to observe objects?

Materials *(per group)*

small pieces of newspaper print and
   colorful magazine photographs
microscope slide
medicine dropper
coverslip
microscope

Procedure

1. Your teacher will instruct you as to the proper use and care of the microscope. Follow all instructions carefully.

2. Obtain a small piece of newspaper print and place it on a clean microscope slide.

3. To make a wet-mount slide, use the medicine dropper to carefully place a drop of water over the newsprint.

4. Carefully lower the coverslip over the newsprint.

5. Place the slide on the stage of the microscope. The newsprint should be facing up and should be in the normal reading position.

6. With the low-power objective in place, focus on a specific letter in the newsprint.

7. Move the slide to the left, then to the right.

8. Looking at the stage and objectives from the side, turn the nosepiece until the high-power objective clicks into place.

9. Using only the fine adjustment knob, bring the letter into focus. Draw what you see.

10. Repeat steps 2 through 9 using magazine samples.

Observations

1. While looking through the microscope, in which direction does the object appear to move when you move the slide to the left? To the right?

2. What is the total magnification of your microscope under low and under high power?

3. What happens to the focus of the objective lens when you switch from low power to high power?

Analysis and Conclusions

1. What conclusion can you draw about the way objects appear when viewed through a microscope?

2. How would you center an object viewed through a microscope when the object is off-center to the left?

3. What is the purpose of the coverslip?

4. **On Your Own** With your teacher's permission, examine samples of hair, various fabrics, and prepared slides through the microscope.

Summarizing Key Concepts

2–1 The Cell Theory

▲ Matthias Schleiden, Theodor Schwann, and Rudolf Virchow were scientists who contributed to the development of the cell theory.

▲ The cell theory states the following: All living things are made of cells; cells are the basic units of structure and function in living things; and living cells come only from other living cells.

2–2 Structure and Function of Cells

▲ The cell wall gives protection and support to plant cells.

▲ The cell membrane regulates the movement of materials into and out of the cell.

▲ The nucleus is the control center of the cell.

▲ Chromosomes are found in the nucleus. Chromosomes direct the production of proteins in the cell and are responsible for cell growth and reproduction.

▲ The endoplasmic reticulum is the site of the manufacture and transport of proteins.

▲ Ribosomes are the protein-making organelles.

▲ Mitochondria, the powerhouses of the cell, provide the energy that cells need to function.

▲ Vacuoles store food, water, and wastes.

▲ Lysosomes contain digestive enzymes, which can be used to break down dead and aging cell parts as well as to digest large food particles in the cell.

▲ Chloroplasts capture energy from the sun and use it to make food for plant cells.

▲ In multicellular organisms, cells are often specialized to perform specific tasks.

▲ The least complex level of organization of living things is the cell.

▲ Cells that are similar in structure and function are often joined together to form tissues.

▲ Groups of different tissues work together as organs.

▲ A group of different organs working together to perform certain functions is known as an organ system.

▲ An organism is an entire living thing that carries out all the life functions.

Reviewing Key Terms

Define each term in a complete sentence.

2–1 The Cell Theory
cell
organelle
cell theory

2–2 Structure and Function of Cells
cell wall
cell membrane
nucleus
chromosome
cytoplasm

endoplasmic reticulum
ribosome
mitochondrion
vacuole
lysosome
chloroplast
division of labor
tissue
organ
organ system

Chapter Review

Content Review

Multiple Choice

Choose the letter of the answer that best completes each statement.

1. The outer covering of an animal cell is the
   a. cell wall.
   b. organelle.
   c. cell membrane.
   d. mitochondria.

2. The control center of the cell is the
   a. cytoplasm.
   b. nucleus.
   c. mitochondria.
   d. nucleolus.

3. Structures involved in the digestive activities of the cell are
   a. lysosomes.
   b. chloroplasts.
   c. nuclear membrane.
   d. endoplasmic reticulum.

4. Protein factories in the cell are known as
   a. mitochondria.
   b. endoplasmic reticulum.
   c. ribosomes.
   d. chloroplasts.

5. Eyes, kidneys, and skin are examples of
   a. tissues.
   b. cells.
   c. organs.
   d. organ systems.

6. The network of passageways that transports proteins throughout the cell is known as the
   a. nuclear membrane.
   b. endoplasmic reticulum.
   c. chloroplast.
   d. ribosome.

7. Food, water, and wastes are stored in
   a. vacuoles.
   b. mitochondria.
   c. ribosomes.
   d. nucleus.

8. The food-making structures in plant cells are called
   a. mitochondria.
   b. chlorophyll.
   c. chromosomes.
   d. chloroplasts.

True or False

If the statement is true, write "true." If it is false, change the underlined word or words to make the statement true.

1. The ability of a cell to maintain a stable internal environment is called <u>homeostasis</u>.
2. The outer covering of <u>animal cells</u> is called the cell wall.
3. Chromosomes are made up of <u>nucleic acids</u>.
4. <u>Vacuoles</u> are the cell's powerhouses.
5. The <u>endoplasmic reticulum</u> is the "brain" of the cell.
6. <u>Tissues</u> make up the second level of organization in living things.
7. Chloroplasts contain <u>cellulose</u>, which is the substance that helps to capture the energy of sunlight.
8. Ribosomes are involved with the production of <u>nucleic acids</u>.

Concept Mapping

Complete the following concept map for Section 2–1. Refer to pages D6–D7 to construct a concept map for the entire chapter.

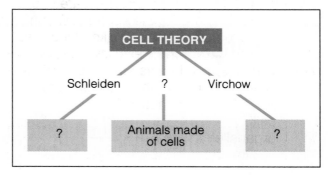

Concept Mastery

Discuss each of the following in a brief paragraph.

1. Based on your knowledge of cells, explain this statement. "One-celled organisms are often more complex than the individual cells of multicellular organisms."
2. Compare a typical plant cell and an animal cell.
3. Describe five functions of a cell and name the structure involved in each function.
4. What were the contributions of Hooke, van Leeuwenhoek, Schleiden, Schwann, and Virchow to our understanding of the cell?
5. Why is it important that the cell membrane be selective in allowing materials into and out of the cell?
6. Explain the relationship between cells, tissues, organs, and organ systems.

Critical Thinking and Problem Solving

Use the skills you have developed in this chapter to answer each of the following.

1. **Classifying organelles** On a sheet of paper, list the various activities of a cell. Next to each activity, give the organelle that is involved in that activity.
2. **Sequencing events** Prepare a time line that illustrates the events that led to the cell theory.
3. **Making predictions** How might the evolution of animals have been affected if animal cells contained a cell wall?
4. **Interpreting graphs** Enzymes are substances that control the speed at which certain reactions in a cell take place. Use the graph to answer the following questions: What two factors are described in the graph? How are these two factors related to enzyme activity?

5. **Making models** To make a model of a cell, dissolve some colorless gelatin in warm water. Pour the gelatin into a rectangular pan (for a plant cell) or a round pan (for an animal cell). Using edible materials that resemble cell structures, place these materials in the gelatin before it begins to gel. On a sheet of paper, develop a key that identifies each cell structure.
6. **Making inferences** Why do scientists believe that mitochondria may have been invaders of early cells?
7. **Making comparisons** Compare the levels of organization in living things to a large factory.
8. **Using the writing process** Write a "Declaration of Independence" for single-celled organisms in which you demonstrate that they are just as complex as multicellular organisms and deserve all the same rights and protections.

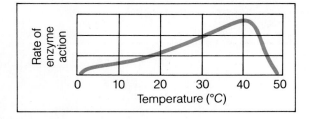

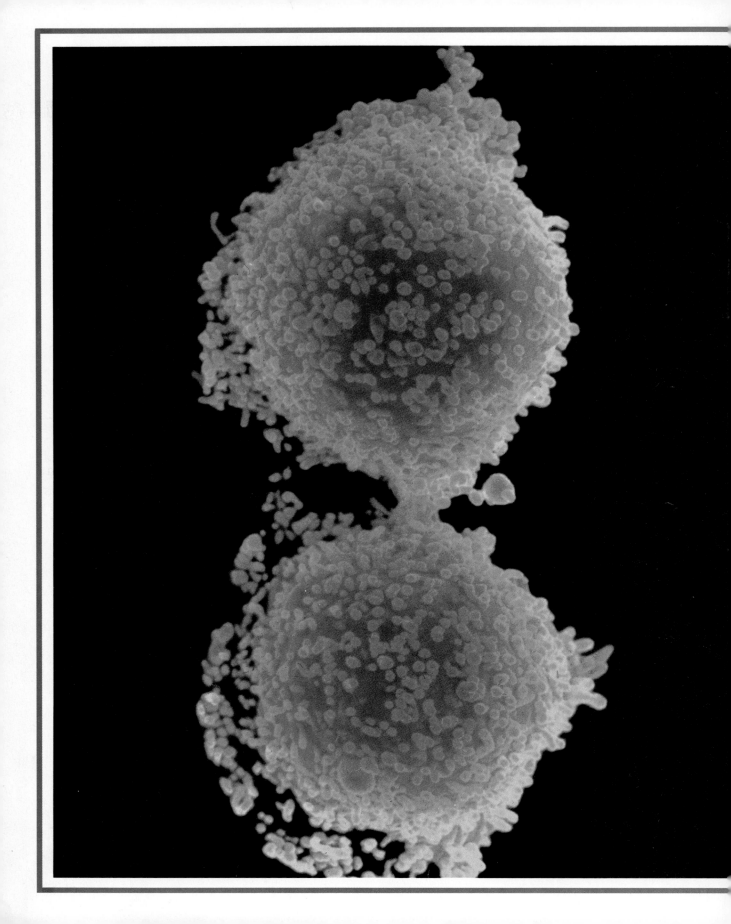

Cell Processes

Guide for Reading

After you read the following sections, you will be able to

3–1 Moving Materials Into and Out of the Cell

- Compare diffusion and osmosis.
- Define and explain the importance of active transport.

3–2 Cell Growth and Division

- Describe the events that occur during cell division.
- Define mitosis.

Trillions of these tiny structures make up your body. Some are in the shape of a rectangle, others are in the shape of a sphere, and still others are spindle-shaped. Some have tails. Others are star-shaped. What are these structures? If you have not already guessed, they are cells.

Cells are the basic units of life. All living things—oak trees, spiders, elephants, and people—are made of cells. Some living things are made of only one cell. Others are made of many cells. But, whether an organism is unicellular or multicellular, its survival depends on the proper functioning of its cell parts.

In Chapter 2 you explored the basic structure of a typical cell and the functions of each cell part. In this chapter you will discover how a cell obtains the raw materials it needs to perform its job. You will also discover how a cell removes its waste products. And you will learn how one cell becomes two cells in a process called cell division.

Journal *Activity*

You and Your World Words often have many meanings. To a biologist, a cell is the basic unit of living things. But to an electrician, the word cell might bring to mind batteries, many of which are made of "dry cells." In your journal, make a list of the different ways the word cell is used. If you have an artistic streak, make a drawing of each definition. You might want to begin with one of your own body cells.

◀ *Notice how the two daughter cells are pulling apart as a cell undergoes cell division.*

3–1 Moving Materials Into and Out of the Cell

Even while you sleep, you need energy to keep you alive. Where does this energy come from? Cells provide it. Although cells cannot make energy, they can change energy from one form to another. Cells obtain energy from their environment and convert it into a usable form. (You will learn more about cell energy in Chapter 4.)

The energy-conversion process in cells is very complex. It involves many chemical reactions. Some reactions break down molecules. Other reactions build new molecules. In Chapter 1 you learned that the sum of all the building-up and breaking-down activities that occur in a living cell is called metabolism.

Metabolism cannot just happen. In order for cells to perform their many functions, the parts of a cell

Figure 3–1 *A cell is like a miniature factory that carries out all the activities necessary to life. Is the factory in the diagram a representation of a plant cell or an animal cell? How do you know?*

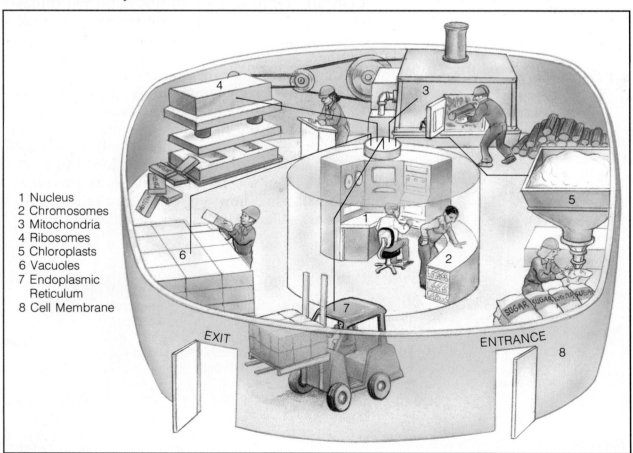

1 Nucleus
2 Chromosomes
3 Mitochondria
4 Ribosomes
5 Chloroplasts
6 Vacuoles
7 Endoplasmic Reticulum
8 Cell Membrane

EXIT

ENTRANCE

need raw materials. And they also need to eliminate poisonous wastes. As you may recall from Chapter 2, the cell membrane forms a barrier between the inside of a cell and the outer environment. So it is a logical assumption (and a correct one) that materials must enter and leave a cell through the cell membrane. **Materials enter and leave a cell by one of three methods: diffusion, osmosis, or active transport.**

Diffusion

Although the cell membrane forms a protective barrier around the cell, it cannot be a total barrier. If it were, nothing could get into or out of the cell. Cell membranes in living things are permeable (PER-mee-uh-buhl) membranes. A permeable membrane allows materials to pass through it. Because it is permeable, the cell membrane allows food molecules, oxygen, water, and other substances to enter or leave the cell. These materials pass through the pores (openings) in the cell membrane.

The driving force behind the movement of many substances into or out of a cell is called **diffusion** (dih-FYOO-zhuhn). **Diffusion is the process by which molecules of a substance move from areas of higher concentration of that substance to areas of lower concentration of that substance.**

Why does diffusion occur? Molecules of all substances are in constant motion, continuously colliding with one another. This motion causes the

Activity Bank

Coming and Going, p. 109

Figure 3–2 *The cell membrane is selective, permitting oxygen and food molecules to enter and waste materials to leave.*

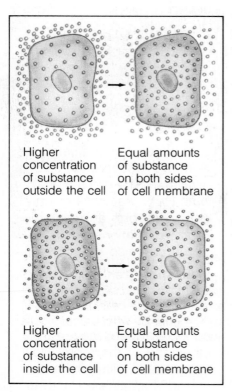

Higher concentration of substance outside the cell

Equal amounts of substance on both sides of cell membrane

Higher concentration of substance inside the cell

Equal amounts of substance on both sides of cell membrane

Figure 3–3 *Diffusion is the movement of molecules of a substance into a cell (top) or out of a cell (bottom). Substances move from places where they are more concentrated to places where they are less concentrated.*

molecules to spread out. The molecules move from an area where there are more of them (higher concentration) to an area where there are fewer of them (lower concentration). See Figure 3–3.

If there are many food molecules outside a cell, for example, some will diffuse through the membrane into the cell. At the same time, waste materials that build up in the cell will diffuse out of the cell.

If substances can move into and out of the cell through the cell membrane, why don't cell organelles and the cytoplasm do likewise? What keeps the ribosomes and mitochondria, for example, from passing out of the cell? And what keeps harmful materials from moving in? The answer is simple but quite elegant. The cell membrane is **selectively permeable**. That is, it permits only certain substances—mainly oxygen, water, and food molecules—to diffuse into the cell. Waste products such as carbon dioxide are allowed to diffuse out of the cell.

Osmosis

Water is the most important substance that passes through the cell membrane. In fact, about 80 percent of the cell is made of water. Water passes through the cell membrane by a special type of diffusion called **osmosis** (ahs-MOH-sihs). Osmosis is the diffusion of water into or out of the cell. During osmosis, water molecules move from a place of higher concentration to a place of lower concentration. This movement keeps the cell from drying out.

Suppose you put a cell into a glass of salt water. The concentration of water outside the cell is lower than the concentration of water inside the cell. This is because there are salt molecules taking up space in the salt water, so there are fewer water molecules. Water leaves the cell, and the cell starts to shrink. If too much water leaves the cell, the cell dries up and dies. Using this information, can you now explain why it is not a wise idea for a person to drink salt water—no matter how thirsty that person is?

ACTIVITY

DISCOVERING

Dissolving Power

Many of the substances that diffuse through the cell membrane are dissolved in the watery fluids surrounding the cell. Do all substances dissolve in water?

■ Using several glass tumblers and common substances found around the home (salt, sugar, starch, flour, baking soda, and so on), determine what substances dissolve in water and what substances do not dissolve in water.

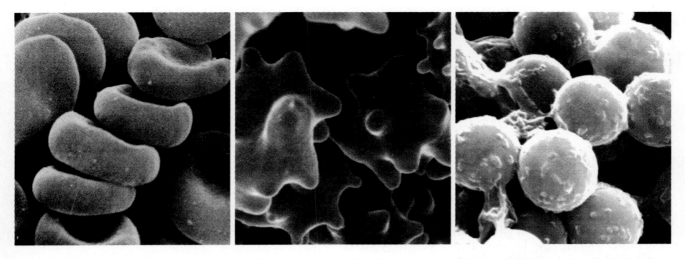

If a cell is placed into a glass of pure, fresh water instead of salt water, just the opposite occurs. Water enters the cell, and the cell swells. This happens because the concentration of water is lower inside the cell than it is outside the cell. As you might imagine, if too much water enters the cell, the cell bursts. Do you remember the cell organelle called the vacuole that you read about in Chapter 2? Vacuoles store food and water for the cell. Some of these vacuoles are contractile vacuoles. As their name suggests, contractile vacuoles can contract, or become smaller. As contractile vacuoles that store water contract, they force excess water out of the cell through the cell membrane.

One of the neat things about diffusion and osmosis is that they do not require energy. That is, the movement of substances across the cell membrane does not require the cell to use up any of its energy reserves. The movement just happens whenever there are unequal concentrations inside and outside a cell. Sometimes, however, the cell must obtain raw materials that cannot diffuse through the cell membrane. At such times, the cell must use some of its available energy to get the materials it requires.

Figure 3–4 *Normal red blood cells (left) will shrink (center) if too much water leaves the cells. If too much water enters the cells, the cells will swell (right). By what process does water move into and out of a cell?*

Figure 3–5 *Notice how the organelles fill a normal plant cell (left). When too much water leaves the cell, the cell contents shrink away from the cell wall (right).*

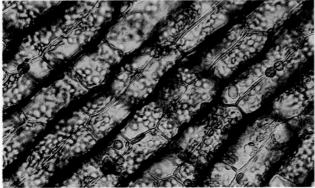

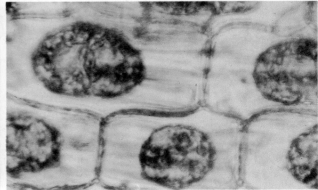

PROBLEM Solving

Shipwrecked!

It had been several days since the Peterson family was marooned on the island. Little did they know that their long-awaited sailing trip would end like this! Things had already been tough, but now they were going to get even tougher. The last of the supplies from their capsized sailboat was almost gone. And they were soon to run out of fresh water.

The Petersons will not last long without water. They need your help. Your job is to find out how they can get the much needed water. But before you tackle that job, see if you can answer these questions.

Can the marooned family simply drink the sea water that they see all around them? Or is there something about sea water that will not quench their thirst? Explain your answer.

Now you are ready to devise a way to provide the Petersons with water. Using any or all of the following items found on the island or taken from the capsized sailboat, try to save the Petersons.

bucket   driftwood
bamboo   coconut
saw   matches
palm leaves

How were you able to save the Peterson family?

Active Transport

As you have just learned, the cell membrane is selectively permeable. It can "select" the materials that will pass into or out of the cell. But what if the cell requires substances that cannot simply diffuse through the cell membrane because the membrane is not permeable to those substances? Or what if the cell membrane is permeable, but the concentration of the substances outside the cell is not high enough to cause diffusion to occur? Is the cell out of luck? Not really. In both cases, the cell can use a process called **active transport** to "carry" the substance into the cell. Unlike diffusion and osmosis, however, active transport requires the cell to use some of its energy reserves.

There are several methods of active transport available to cells. In the most common method, special transport molecules in the cell membrane actually pick up the substance outside the cell and pull it through the cell membrane. Some substances needed by the cell that are carried in this manner are calcium, potassium, and sodium. Active transport is also used to eliminate substances inside the cell that cannot pass through the cell membrane by diffusion. Regardless of whether materials are passing into or out of the cell, the important thing to remember is that active transport requires energy. Unlike diffusion and osmosis, active transport does not just happen on its own.

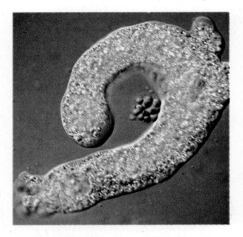

Figure 3–6 *Amebas use a form of active transport in which a large food particle is surrounded by pockets of the cell membrane. Once the food is surrounded, the pocket breaks away from the cell membrane and forms a vacuole within the ameba.*

3–1 Section Review

1. How do materials move into and out of a cell?
2. Compare diffusion, osmosis, and active transport.
3. Red blood cells require potassium to function properly. The concentration of potassium molecules inside a red blood cell, however, is usually higher than the concentration of potassium in the blood. Explain how potassium molecules are likely to enter a red blood cell.

Connection—*Science and Technology*
4. Waste products in the blood are filtered out by the kidneys. In the past, kidney failure always led to death because of the buildup of poisonous wastes in the blood. Today, however, doctors use dialysis machines to filter wastes out of the blood. Dialysis machines use artificial membranes. Are the membranes in a dialysis machine permeable or selectively permeable? Explain.

ACTIVITY

DISCOVERING

Eggs-periment

Carefully peel two hard-boiled eggs. Obtain a string, a metric ruler, 2 beakers, some water, and salt.

■ Using these materials, design and perform an experiment in which you show that during osmosis, water molecules move from an area of higher concentration to an area of lower concentration. (For best results, you should perform your experiment over a period of about five days.)

ACTIVITY

CALCULATING

How Many Cells?

Suppose a cell divides once a day. How many cells will there be in a week? A month? A year?

3–2 Cell Growth and Division

You learned in Chapter 1 that growth is one of the characteristics of living things. Growth is usually a fairly obvious characteristic. A human infant, for example, clearly shows growth as it passes through childhood and into adulthood. The growth of a red-wood seedling into a tree is also quite apparent.

It might seem that the easiest way for an organism to grow is for its cells to get larger and larger. In the case of our previous example, the cells in an infant might get bigger and bigger as that infant developed into an adult. The fact is, however, that cells do not grow larger and larger in this manner. The cells in a human infant are about the same size as the cells in an adult human—an adult just has a lot more cells than an infant does!

Limits on Cell Growth

Why don't cells get bigger and bigger as an organism grows? The answer has to do with the transportation of materials into and out of a cell. If a cell continued to grow larger and larger, at some point the cell membrane would not be able to handle the flow of materials passing through it. That is, the amount of raw materials needed by the larger cell would not enter the cell fast enough. The amount of wastes produced by the larger cell could not leave the cell fast enough. The larger cell would then die.

Figure 3–7 *The adult tiger does not have larger cells than its cub, just more of them. Why can't cells grow larger and larger and larger?*

Cell Division

In order for the total number of cells to increase and for an organism to grow, the cells must undergo **cell division**. During cell division, one cell divides into two cells. Each new cell, called a daughter cell, is identical to the other and to the parent cell.

If a parent cell—a skin cell, leaf cell, or bone cell, for example—is to produce two identical daughter cells, then the exact contents of its nucleus must go into the nucleus of each new daughter cell. Recall that the chromosomes, which contain the blueprints of life, are located in the nucleus. If a parent cell simply splits in half, each daughter cell will get only half the contents of the nucleus—only half the chromosomes of the parent cell.

Fortunately, this does not happen. To understand why not, you must know about the process of cell division in more detail. **Cell division occurs in a series of stages, or phases.** Each has a scientific name. It is not important that you memorize the scientific name for each phase. But it is important that you understand the nature of cell division and how a parent cell divides into two daughter cells.

PHASE 1: CHROMOSOMES ARE COPIED   During the first phase of cell division, which is called interphase, the cell is performing its life functions, but it is not actually dividing. If you were to observe the nucleus during this phase, you would not be able to see the rodlike chromosomes. Instead, the chromosomes would appear as threadlike coils called **chromatin**. In animal cells, two structures called centrioles (SEHN-tree-ohlz) can be seen outside the nucleus. The centrioles play a part in cell division. Most plant cells do not have centrioles.

Near the end of phase 1, the process of cell division begins. At this time, all the chromosomes (which still appear as threadlike coils of chromatin) are duplicated. That is, a copy of each chromosome is produced. As a result, the normal chromosome number in the cell doubles. Each chromosome and its sister chromosome (its copy) are attached at an area called the centromere. At this time the sister chromosomes are called chromatids.

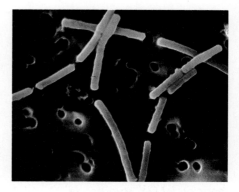

Figure 3–8 *Although bacteria do not contain a nucleus, their chromosomes still must double before the bacterial cells can divide into two identical daughter cells.*

Figure 3–9 *During phase 1 of cell division, or interphase, the cell is performing the normal processes of life.*

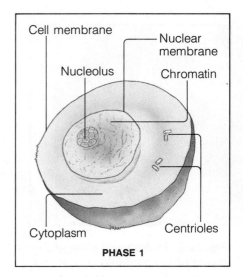

Cell membrane

Nucleolus

Nuclear membrane

Chromatin

Cytoplasm

Centrioles

PHASE 1

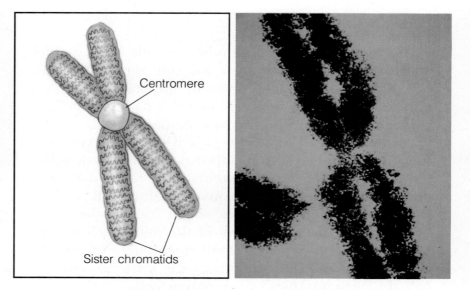

Figure 3–10 *The diagram of a chromosome shows that it consists of two chromatids attached by a centromere. A human chromosome is shown as it appears through an electron microscope.*

Centromere

Sister chromatids

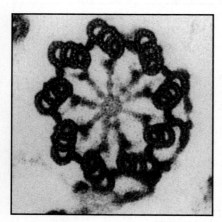

Figure 3–11 *Notice the ringlike shape of the centriole in this electron micrograph. What role do centrioles play in mitosis?*

PHASE 2: MITOSIS BEGINS It is during the second phase, which is called prophase, that cell division really gets going. At this point, the process of cell division is called **mitosis** (migh-TOH-sihs). **Mitosis is the process by which the nucleus of a cell divides into two nuclei and the formation of two new daughter cells begins.**

During phase 2, a number of important events occur. The threadlike chromatin in the nucleus begins to shorten and form the familiar rodlike chromosomes. Each chromosome, which appears as two identical chromatids attached at the centromere, is easily seen under a microscope. Around this time the two centrioles (in animals cells) begin to move to opposite ends of the cell. In addition, a meshlike spindle begins to develop between the two centrioles, forming a "bridge" between the opposite ends of the cell. (Although plant cells do not contain centrioles, a spindle still forms in the cell at this time.) Near the end of phase 2 of cell division, the nuclear membrane surrounding the nucleus begins to break down. At the same time, the nucleolus in the nucleus disappears.

PHASE 3: CHROMOSOMES ATTACH TO THE SPINDLE During phase 3 of cell division, which is called metaphase, the chromosomes begin to attach to the meshlike spindle that runs from end to end in the cell. In phase 3, the chromosomes are attached to the spindle by the centromere, which still connects each chromatid to its identical sister chromatid.

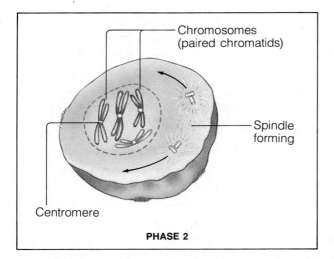

Chromosomes
(paired chromatids)

Spindle
forming

Centromere

PHASE 2

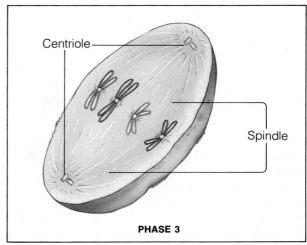

Centriole

Spindle

PHASE 3

PHASE 4: CHROMOSOMES BEGIN TO SEPARATE

During phase 4 of cell division, which is called anaphase, the centromere, which holds two sister chromatids together, splits. This allows the sister chromatids to separate from each other. One chromatid from each pair of sister chromatids begins to move toward one end of the cell along the spindle. The other chromatid of the pair begins to move toward the other end of the cell along the spindle. In other words, the chromatids in a pair separate so that each sister chromatid goes to an opposite end of the cell. At this point, the chromatids are again called chromosomes.

PHASE 5: TWO NEW NUCLEI FORM

During phase 5 of cell division, which is called telophase, the chromosomes begin to uncoil and lose their rodlike

Figure 3–12 *During phase 2, or prophase, the process of mitosis begins and the chromatin condenses and takes on the familiar rodlike chromosome shape. In phase 3, or metaphase, the chromosomes attach to the spindle.*

Figure 3–13 *In phase 4, or anaphase, the centromeres holding sister chromatids split, allowing each chromatid to become an individual chromosome. By phase 5, or telophase, the chromosomes have segregated to opposite ends of the cell and the nuclear membrane begins to re-form.*

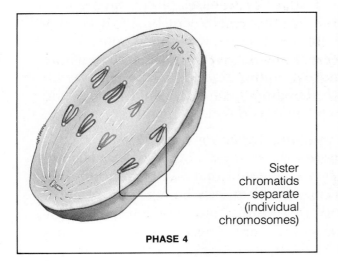

Sister
chromatids
separate
(individual
chromosomes)

PHASE 4

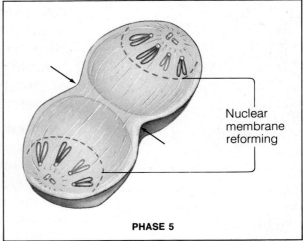

Nuclear
membrane
reforming

PHASE 5

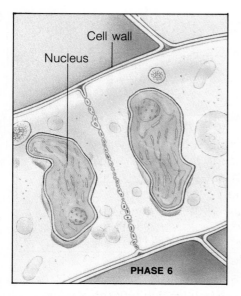

Figure 3–14 *Cytokinesis is the division of the cytoplasm and its contents into two individual daughter cells. In plant cells, as shown here, the cytoplasm is divided by a cell plate, which will become the new cell membrane.*

appearance. The chromosomes again appear as chromatin and cannot easily be distinguished under a microscope. A nuclear membrane forms around the chromatin at each end of the cell. In each nucleus, a nucleolus reappears. At this point, the process of mitosis is complete. But cell division still has one more phase to go before it is over.

PHASE 6: TWO DAUGHTER CELLS FORM The sixth and final phase of cell division, which is called cytokinesis (sigh-toh-kuh-NEE-suhs), involves the division of the cytoplasm in the cell. During this phase, the membrane surrounding the cell begins to move inward until the cytoplasm is pinched into two nearly equal parts. Each part contains a nucleus with identical chromosomes. Soon the cell membrane completes its job and two new daughter cells form. In plant cells, a cell wall also forms around each daughter cell.

The process of cell division is now over. Two new daughter cells have formed, each having the same number and kind of chromosomes as the original parent cell.

In multicellular organisms, cell division takes place millions of times as an organism grows and develops. Cell division also occurs to replace dead or injured cells. Where do you think cell division might occur in your body?

3–2 Section Review

1. Briefly describe the phases of cell division.
2. An adult has more cells, not bigger cells, than an infant has. Explain why this is so.
3. During mitosis, the sister chromatids separate and move to opposite ends of the cell. Describe the two daughter cells that would form if the sister chromatids did not separate.

Critical Thinking—*Applying Concepts*
4. During sexual reproduction, two sex cells unite to form a single cell that will become the new organism. Would mitosis be an appropriate method for forming sex cells? Explain your answer.

I Feel Dizzy, Oh So Dizzy!

Travel is something we all take for granted in this high-speed age of ours. But think about this. The first Paris to New York nonstop flight in an airplane took 33 hours and 30 minutes, and occurred in 1930. The airplanes that crowd our skies today and that can fly you to Paris in about 4 hours are a comparatively new invention.

What happened before airplanes made transatlantic crossings so popular with millions of travelers? Transatlantic crossings were made by ship. For the very wealthy, these crossings were luxurious and opulent. Fine foods, service, and many changes of clothing each day made life aboard ship pleasurable. For many others—poor immigrants, for example—leaving Europe for a new life in a new land, the unpleasant crossings were made bearable by the hope of a better future.

But for all people, rich and poor alike, one of the most unpleasant factors in an ocean crossing was *seasickness*. Sea-sickness is caused by the movement of a boat as it plows through the waves. For even though ocean liners are huge and powerful, the ocean is larger and more powerful still. Ocean waves can toss even the largest ship about with abandon. The constant motion of the ocean causes some people to feel nauseated. And after several days trapped on a ship this feeling can be very unpleasant indeed.

Today there is an easy treatment for seasickness, and one that uses a technique you learned about in the chapter—diffusion. Physicians place a small bandagelike patch behind a person's ear. The patch contains a special drug called scopolamine, which enters the body by diffusing through the skin at a constant rate. This method of treatment is successful in most people. This technique is being used to administer other medicines as well. In the future these patches may become even more commonplace as treatments for other diseases are developed.

Laboratory Investigation

Observing Mitosis

Problem

How do the phases of mitosis appear under a microscope?

Materials *(per group)*

prepared slides of mitosis in animal and plant cells
microscope

Procedure 🧪

1. Begin your investigation by observing prepared slides of mitosis in animal cells.

2. Examine the slides in the order that corresponds to the phases of mitosis.

3. Draw and label what you observe on each slide.

4. Now observe prepared slides of mitosis in plant cells in the order that corresponds to the phases of mitosis.

5. Again draw and label what you observe.

Observations

Place your drawings of each phase of mitosis in animal and plant cells side by side.

Analysis and Conclusions

1. Based only on your observations, compare mitosis in animal and plant cells.

2. Were there events in mitosis you read about in this chapter that you could not observe under the microscope? If so, which events were they?

3. Mitosis is only one part of the cell division process. Did any of your slides show other phases of cell division? If so, what were they?

4. **On Your Own** Using materials of your choice, construct a three-dimensional model of mitosis in a plant cell or in an animal cell.

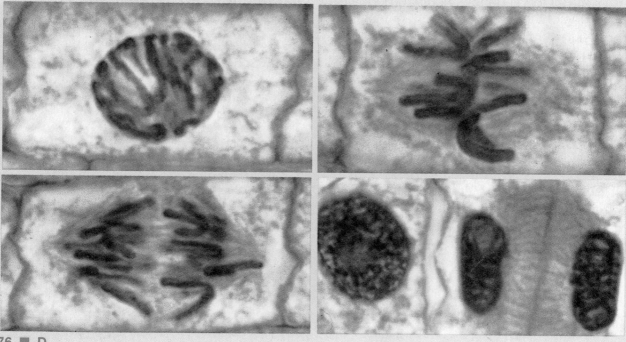

Summarizing Key Concepts

3–1 Moving Materials Into and Out of the Cell

▲ Materials pass into and out of the cell through pores in the cell membrane. The cell membrane is selectively permeable, allowing some materials to pass through and preventing others from entering or leaving the cell.

▲ Materials pass through the cell membrane by one of three processes: diffusion, osmosis, or active transport.

▲ Diffusion is the process by which molecules of a substance move from areas of higher concentration of that substance to areas of lower concentration of that substance.

▲ Osmosis is the diffusion of water.

▲ Neither diffusion nor osmosis requires the cell to expend energy.

▲ Materials can move into the cell through active transport, which requires the cell to expend energy.

3–2 Cell Growth and Division

▲ The cells of an organism increase in number through cell division. During cell division, one cell divides into two daughter cells, each of which is identical to the original parent cell.

▲ Cell division occurs in a series of phases, or stages. In phase 1 of cell division (interphase), the number of chromosomes in the nucleus doubles.

▲ Phase 2 of cell division (prophase) marks the beginning of the process called mitosis. During mitosis, the nucleus of a cell divides into two nuclei, and the formation of two new daughter cells begins.

▲ During phase 2 of cell division, the nuclear membrane breaks down. In addition, a mesh-like spindle running from one end of the cell to the other end forms.

▲ In phase 3 of cell division (metaphase), the chromosomes, which now consist of paired, sister chromatids, attach to the spindle at their centromere.

▲ In phase 4 of cell division (anaphase), the centromere holding the sister chromatids splits. One chromosome from each pair of sister chromatids moves along the spindle toward an opposite end of the cell.

▲ Phase 5 of cell division (telophase) marks the end of the process called mitosis. During this phase, a nuclear membrane forms around each set of chromosomes at the ends of the cell.

▲ Phase 6, the final phase of cell division (cytokinesis), starts when the cell membrane begins to pinch in, forming two new daughter cells. Each daughter cell contains the same number and kind of chromosomes as the original parent cell.

Reviewing Key Terms

Define each term in a complete sentence.

3–1 Moving Materials Into and Out of the Cell

diffusion
selectively permeable
osmosis
active transport

3–2 Cell Growth and Division

cell division
chromatin
mitosis

Chapter Review

Content Review

Multiple Choice

Choose the letter of the answer that best completes each statement.

1. The cell membrane
   a. provides protection for the cell.
   b. keeps the cell's inner contents together.
   c. regulates the movement of materials into and out of the cell.
   d. does all of these.
2. Water usually moves through a cell membrane from an area of
   a. lesser concentration to an area of higher concentration.
   b. equal concentration to an area of equal concentration.
   c. higher concentration to an area of lesser concentration.
   d. none of these.
3. The total number of cells in an organism increases as a result of
   a. respiration.   c. osmosis.
   b. cell division.   d. homeostasis.
4. Chromatids are held together by a
   a. spindle.   c. centromere.
   b. centriole.   d. cell membrane.

5. A cell has 12 chromosomes. How many chromosomes will each daughter cell have after cell division?
   a. 24   c. 12
   b. 6   d. 48
6. A structure found in animal cell division that is not usually found in plant cell division is the
   a. spindle.   c. cell wall.
   b. centromere.   d. centriole.
7. A membrane that allows all materials to pass through is
   a. selectively permeable.
   b. impermeable.
   c. permeable.
   d. none of these.
8. The process in which the nucleus of a cell divides into two nuclei and the formation of two new daughter cells begins is called
   a. osmosis.
   b. mitosis.
   c. diffusion.
   d. spindle formation.

True or False

If the statement is true, write "true." If it is false, change the underlined word or words to make the statement true.

1. As an organism grows, its cells increase in <u>size</u>.
2. As a result of mitosis, a cell having 20 chromosomes gives rise to two cells, each of which contains <u>20</u> chromosomes.
3. During phase 3 of cell division (metaphase), chromosomes attach to the <u>spindle</u>.
4. The movement of water through a cell membrane is called <u>diffusion</u>.
5. <u>Meiosis</u> begins during phase 2 of cell division (prophase).

Concept Mapping

Complete the following concept map for Section 3–1. Refer to pages D6–D7 to construct a concept map for the entire chapter.

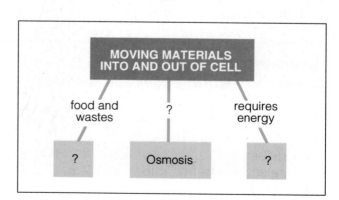

Concept Mastery

Discuss each of the following in a brief paragraph.

1. Why is it vital that a cell membrane be selectively permeable?
2. How do living things grow? What is the main factor that limits cell growth?
3. Describe and compare chromosomes, chromatin, chromatids, and centromeres.
4. Describe and compare diffusion, osmosis, and active transport.
5. How does the process of cell division ensure that daughter cells will be identical to the parent cell?

Critical Thinking and Problem Solving

Use the skills you have developed in this chapter to answer each of the following.

1. **Relating facts** Describe the events that occur during each phase of cell division.
2. **Relating cause and effect** Your favorite plant has begun to wilt. You feel the soil in its pot and find that it is very dry. You water your plant and about 20 minutes later you discover that the plant is standing up straight again. How does this observation relate to osmosis?

3. **Drawing conclusions** In a famous science fiction movie called *The Blob,* a giant, amebalike cell terrorizes a community, eating many of its residents. What basic fact about cells did the moviemakers disregard for the sake of drama?
4. **Making models** Construct a three-dimensional model of a phase of cell division in either a plant cell or an animal cell. Label the phase you are modeling and provide a key to explain what is happening in the cell.
5. **Developing a hypothesis** Fill one beaker with tap water at room temperature, another beaker with ice water, and a third beaker with hot water. Add equal amounts of food coloring to each beaker. In which beaker does the food coloring diffuse the fastest? The slowest? What is the variable in this experiment? Which beaker is the control? State a hypothesis for this experiment.
6. **Applying concepts** Is mitosis occurring in your body at this moment? If so, where? Explain your answer.
7. **Using the writing process** Write a short story called "The Day Diffusion Went Backwards."

Cell Energy

Guide for Reading

After your read the following sections, you will be able to

4–1 Photosynthesis: Capturing and Converting Sunlight

- Describe the process of photosynthesis.
- Identify the substances plants require to perform photosynthesis.
- Compare autotrophs and heterotrophs.

4–2 Respiration: Using the Energy in Food

- Describe the process of respiration.
- Compare the general equations for respiration and photosynthesis.

Take a moment to study the photograph on the opposite page. The scene seems tranquil. Perhaps a slight breeze is blowing across the tops of the flowers. There might even be a rabbit or two scurrying about on the meadow floor. Beneath the surface of the ground, worms are burrowing through the soil in search of food. But all in all, not much is happening—or is it?

Actually, a lot is going on. And it all starts with the sun. Each morning, as the meadow is bathed in sunlight, a new day begins in the struggle of living things to obtain energy. Sunlight is the source of energy for just about all living things on planet Earth. Without the sun there would be no life. Why? The cells of all living things require energy. Without a constant supply of energy, cells quickly die.

Energy is the ability to do work. And there is certainly plenty of work to be done by cells. In this chapter, you will learn about some of the ways cells capture, store, and use energy. When you have completed the chapter, perhaps you will have a new appreciation of Earth's brightest neighbor in the sky—the sun!

Journal *Activity*

You and Your World Our sun is an unusual star. While most stars come in pairs, the Earth's sun sits alone in space with only its planetary neighbors. Imagine a world with two suns—a world in which there is constant daylight. In your journal, describe how you would feel if night never fell and the sun shone all the time.

◄ *These flowers, trees, and grasses are all involved in capturing energy from sunlight and transforming it into energy that can be used by all living things.*

4–1 Photosynthesis: Capturing and Converting Sunlight

Have you ever wondered why you eat? Sure, food tastes good, and it keeps your stomach from growling. But there must be a more important reason you ingest (take in) food.

Some of the foods you eat are needed as raw materials for building new cells and repairing damaged or worn-out cells. But most of the foods you eat are used as sources of energy. That energy is locked up in the chemical bonds in the foods. Chemical bonds link atoms together. Remember, atoms are the building blocks of all forms of matter. Inside your body, the energy in those chemical bonds can be released and used by your cells. You will learn how that happens in the next section of this chapter.

There is plenty of energy on Earth. Raging winds carry energy across the globe. Volcanoes release huge amounts of energy as they erupt. The energy of an earthquake is powerful enough to level entire cities. Fuels such as wood, gasoline, oil, and coal provide energy to homes and factories. Sometimes it may seem as if the Earth is awash in energy. But none of that energy is available to your cells. That is, your body has no way of converting the energy of a volcano or a lump of coal into a form it can use.

Figure 4–1 *Like other animals, people get their energy from the foods they eat, whether it be a Mexican banquet, Japanese sushi, or Italian spaghetti.*

Food is the only source of energy your body has. But where does the energy in food come from? As you have read in the opening of this chapter, the ultimate source of energy for living things is the sun. Can you use sunlight to make food? You know the answer to that question already. You cannot—but some organisms can. These organisms are mainly green plants. So your exploration into how cells obtain energy must begin with the energy-converting, food-making process called **photosynthesis** that takes place in green plants. (The prefix *photo-* means light, and the root word *-synthesis* means putting together. So the term photosynthesis means using light to make food.)

Photosynthesis: Capturing Energy

Within the leaves and other green parts of plants, the most important manufacturing process on Earth—the process of photosynthesis—takes place. You may recall from Chapter 2 that plant cells contain organelles called chloroplasts. Within chloroplasts is the green pigment known as chlorophyll. (Chlorophyll is what makes green plants green.) Photosynthesis takes place in the chloroplasts.

The first step in photosynthesis occurs when energy in the form of sunlight is captured by the chlorophyll in a chloroplast. The exact nature of this capturing process is quite complicated. Although you

The Colors of Sunlight

Sunlight appears to be white. In fact, it is called white light. But is white light actually white?

■ Using a prism and sunlight, determine if the the light from the sun is white or if it is made up of different colors that combine to form white light.

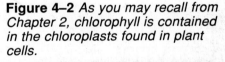

Figure 4–2 *As you may recall from Chapter 2, chlorophyll is contained in the chloroplasts found in plant cells.*

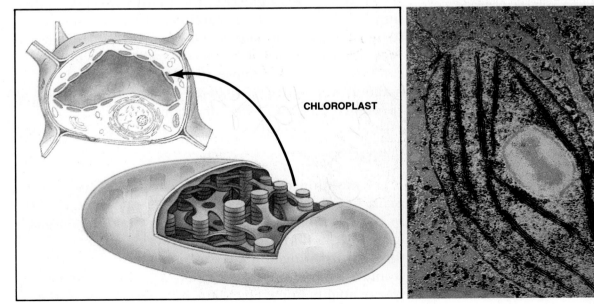

CHLOROPLAST

Figure 4–3 *Plants are not the only organisms that contain chlorophyll and perform photosynthesis. This photosynthetic green euglena is a member of the protist kingdom.*

do not need to know the detailed chemistry of photosynthesis, you do need to understand that green plants (as well as other organisms that contain chlorophyll) can capture energy from sunlight. As you might expect, this energy-capturing process happens only when the sun is shining.

Photosynthesis: Converting Energy

What happens once the chlorophyll has captured some of the sun's energy? **The second step in photosynthesis occurs when the energy that has been captured is converted into the energy found in food.** In other words, the radiant energy of sunlight is converted into the chemical energy of food. This chemical energy is locked into the bonds between the atoms that make up food. A conversion of energy simply means that energy in one form is changed into energy in another form. No energy is lost or gained in a conversion. Only its form is changed. The energy-converting, second step of photosynthesis can occur in the light or in the dark. Only the first step of capturing energy requires sunlight.

The food produced by photosynthesis is a simple sugar called glucose. Glucose is a carbohydrate, or a compound made of carbon, hydrogen, and oxygen atoms. It is within the bonds holding carbon, hydrogen, and oxygen atoms together that the sun's energy is stored. From glucose, the plant can make other simple sugars as well as starches, which are long strings of sugars linked together in a chain. Both sugars and starches are carbohydrates. The plant can also combine glucose with other chemicals to make all the other substances a living thing needs (proteins, fats, and oils, for example).

Requirements for Photosynthesis

In addition to sunlight and chlorophyll, a green plant needs carbon dioxide (CO_2) and water (H_2O) to perform photosynthesis. Carbon dioxide is a gas found in the atmosphere. It enters the plant through openings on the surface of the leaf called **stomata** (STOH-muh-tuh; singular: stoma). Water, as you probably know, is taken in by the roots of the

ACTIVITY

WRITING

The Calvin Cycle

In 1946, Melvin Calvin, a scientist at the University of California at Berkeley, discovered that some stages of photosynthesis can occur in the dark. He called these stages the dark reactions.

Using reference materials from the library, write a brief report on Melvin Calvin's discovery of the Calvin cycle.

plant and transported up the stem to the leaf. So in the leaf you will find carbon dioxide, water, and chlorophyll in the chloroplasts. All the plant needs now is a sunny day.

Although photosynthesis is a complex process, it can be described by both a word equation and a chemical equation. Both equations show the raw materials for the process on the left and the products on the right. An arrow, which is read as "yields," connects the raw materials to the products. Any substances or conditions that are required for the process to occur but that do not actually get changed in the process are written above and below the arrow:

$$\text{Carbon Dioxide} + \text{Water} \xrightarrow[\text{Chlorophyll}]{\text{Sunlight}} \text{Glucose} + \text{Oxygen}$$

or

$$6CO_2 + 6H_2O \xrightarrow[\text{Chlorophyll}]{\text{Sunlight}} C_6H_{12}O_6 + 6O_2$$

According to the word equation, carbon dioxide and water in the presence of sunlight and chlorophyll are converted into glucose (food) and oxygen. The chemical equation uses symbols and formulas to indicate the substances and their actual amounts that are involved in photosynthesis. So 6 molecules of carbon dioxide combine with 6 molecules of water to yield 1 molecule of glucose and 6 molecules of oxygen.

There is something extremely important about the equation for photosynthesis. Did you notice it? Food, in the form of glucose, is not the only product. Another product of photosynthesis is oxygen. Oxygen, however, is considered a waste product of the process. Oxygen leaves the plant through the stomata, the same route through which carbon dioxide entered the plant.

Why is the waste product oxygen so important? If you are not quite sure, take a deep breath and think again. That's right: It is the oxygen produced by plants during photosynthesis that almost all living things, including humans, require for survival. Without photosynthesis, there would be little or no oxygen in the atmosphere. (Recall from Chapter 1 the importance of oxygen in the atmosphere to the evolution of living things.)

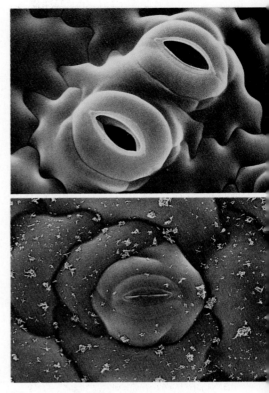

Figure 4–4 *When the stomata on the surface of a leaf are open, carbon dioxide can enter the plant (top). What happens when the stomata are closed (bottom)?*

ACTIVITY

DISCOVERING

Do Plants Breathe?

Obtain a common plant such as a *Coleus*. Cover both the tops and bottoms of half the plant's leaves with petroleum jelly. Place the plant on a sunny windowsill for a week or more.

■ Based on your observations, do plants need to breathe?

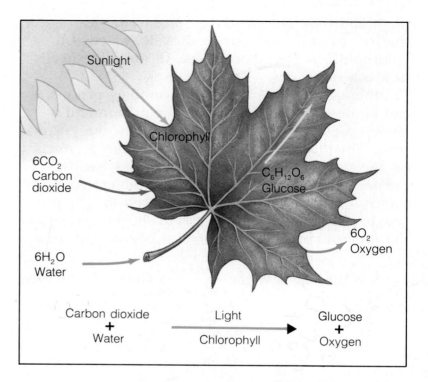

Figure 4–5 *In photosynthesis, the energy of sunlight captured in chlorophyll is used to convert carbon dioxide and water into glucose (food) and oxygen.*

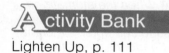

ctivity Bank

Lighten Up, p. 111

Figure 4–6 *This New Hampshire scene reminds us that the leaves of many plants contain pigments other than green chlorophyll. These pigments are hidden until the chlorophyll breaks down during the autumn season.*

Because oxygen is given off as a waste product, you might think that plants do not require it. Not so. Just as do almost all other living things, green plants need oxygen in order to release the energy in food. You will read about how organisms use oxygen in the next section of this chapter.

Autotrophs and Heterotrophs

As you have just learned, green plants and other organisms with chlorophyll can produce their own food during photosynthesis. Green plants are called **autotrophs** (AWT-oh-trohfs), or organisms that can produce their own food.

Animals obtain food by eating green plants or by eating other animals that eat green plants. Organisms that cannot produce their own food and thus must eat other organisms to obtain energy are called **heterotrophs** (HED-uhr-oh-trohfs). Are you an autotroph or a heterotroph?

Figure 4–7 *Green plants are autotrophs, or organisms that produce their own food. Animals are heterotrophs and must obtain their food directly from green plants, as this bushcricket does, or indirectly from animals that eat green plants, as does this barn owl.*

CAREERS

Cell Biologist

How do scientists know so much about cell parts and processes? To a large degree, their knowledge (and ours) of cells has resulted from the work of **cell biologists**. Cell biologists study the structure of cells in all manner of living things. They examine different parts of cells to learn how each part functions and how it is affected by chemical and physical factors.

Cell biologists usually have a medical degree or PhD degree. They work in hospitals, universities, and research laboratories. For more information on this interesting career in biology, contact the American Society for Cell Biology, 9650 Rockville Pike, Bethesda, MD 20814.

4–1 Section Review

1. Define photosynthesis and explain its importance to all living things.
2. What substances does a plant require to perform photosynthesis?
3. Why is oyxgen an important waste product of photosynthesis?

Critical Thinking—*Applying Concepts*

4. Although sunlight appears white, it is actually made up of many different colors of light: red, orange, yellow, green, blue, and violet. Design an experiment to determine if chlorophyll can use all the colors of sunlight or if only some colors are absorbed and used by chlorophyll.

Guide for Reading

Focus on this question as you read.

▶ How does respiration enable organisms to use the energy in food?

4–2 Respiration: Using the Energy in Food

Green plants (autotrophs) make their own food. Animals (heterotrophs) eat green plants or other animals that eat green plants. Either way, organisms manage to get food into their body. But how do organisms use the food? That is, how do organisms get the energy they vitally need from the food?

You have learned that energy is stored in the bonds that link together atoms in foods, such as glucose. You can think of the energy stored in food as a savings account. You eat food to add to your energy savings account. When your cells need energy, they make a withdrawal from the energy savings account.

Aerobic Respiration

In Chapter 2 you read about the powerhouses of the cell, the organelles called mitochondria. It is in the mitochondria that the process of **respiration** occurs. **Respiration is a process in which simple food substances such as glucose are broken down, and the energy they contain is released.** Because living things need a continuous supply of energy, respiration is performed constantly in the cells of all living things.

Figure 4–8 *Mitochondria are the powerhouses of the cell, or the sites where cellular respiration occurs. What substance is combined with food in the mitochondria during respiration?*

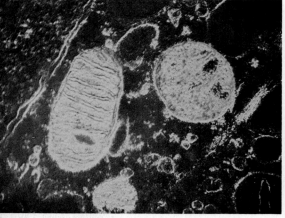

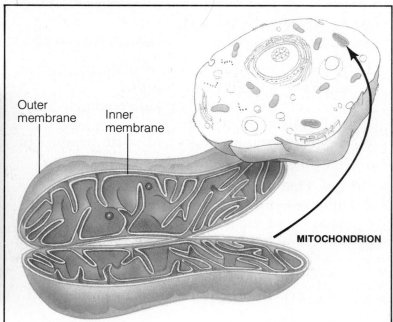

Outer membrane

Inner membrane

MITOCHONDRION

Aerobic respiration, as its name implies, requires oxygen. (The prefix *aero-* means air. Oxygen is one of the main components of air.)

In aerobic respiration, food enters the mitochondria. The food is broken down when it combines with oxygen. During this process, water and carbon dioxide are produced as waste products.

The energy released from food during respiration is not always used right away by the cell. Often, it is stored in a chemical compound called **ATP**. Basically, ATP is a chemical substance that can be used to store energy. When a cell needs energy, it breaks down the ATP and uses the energy released. When a cell has an adequate supply of energy, it keeps the ATP in reserve until it is needed.

Although respiration occurs in a series of complex steps, the overall process can be written as a word equation and as a chemical equation, as follows:

Glucose + Oxygen $\longrightarrow$ **Carbon Dioxide + Water + ATP (energy)**

or

$$C_6H_{12}O_6 + 6O_2 \longrightarrow 6CO_2 + 6H_2O + ATP$$

Notice anything familiar about the equation for respiration? You are quite right if you said that it is the opposite of the equation for photosynthesis. This is an important point to remember. During photosynthesis, carbon dioxide and water are used to produce glucose and oxygen. During respiration, glucose and oxygen are used to produce carbon dioxide and water. Thus, in a stable environment both oxygen and carbon dioxide can be cycled so that both gases can be used by living things over and over again—which is just what a cycle means.

Earlier we said that although plants release oxygen during photosynthesis, they still need oxygen to survive. Now you can understand why. In order to get energy from the foods they produce during photosynthesis, plants require oxygen. In order to get energy from the food you ingest, you require

ACTIVITY READING

The Green Planet

If you have ever complained about doing chores such as mowing the lawn, you may want to read *Grass* by Sherri Tepper. In this science fiction story, the author writes about an entire planet covered with grass. Read the book and discover some of the unusual animals that live on the green planet.

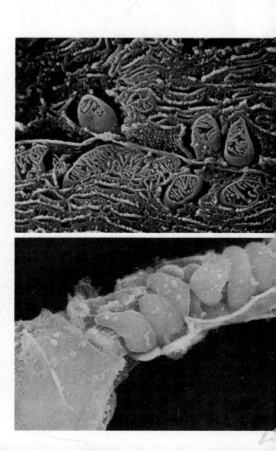

Figure 4–9 *The mitochondria in this thinly sliced human cell appear orange. How would you describe the inner structure of the mitochondria? A sperm cell carries its mitochondria in its tail, which is released before the sperm cell unites with an egg cell. Why do sperm need a plentiful supply of mitochondria?*

oxygen. Where do you get the oxygen you need for respiration? How do you get rid of carbon dioxide? (*Hint:* You don't have stomata on the surface of your skin!)

Fermentation: Anaerobic Respiration

Some organisms are anaerobic, which means they do not need oxygen to survive. But all organisms need energy. So how do anaerobic organisms get their energy? The answer is that anaerobic organisms undergo a different type of respiration, appropriately called anaerobic respiration. Scientists use the term **fermentation** when they speak of anaerobic respiration. Keep in mind that the amount of energy released during fermentation is much less than the amount of energy released during aerobic respiration. That is, aerobic respiration is a much more efficient process than fermentation (anaerobic respiration).

Yeasts are one type of organism that utilizes fermentation. In yeast cells, glucose is broken down. The products of this breaking-down process are carbon dioxide, alcohol, and ATP. These products are important to bakers and brewers. Carbon dioxide produced by yeast causes dough to rise, and it forms the air spaces you see in bread. Carbon dioxide is also the source of bubbles in alcoholic beverages such as beer and wine.

Figure 4–10 *During the production of bread, yeast cells produce carbon dioxide as part of the fermentation process. The carbon dioxide causes the bread to rise and forms the air spaces in the bread. During strenuous exercise, muscles may not get an adequate supply of oxygen. At such times, the muscles use a form of fermentation that produces lactic acid. The buildup of lactic acid causes the painful, burning sensation all athletes have experienced. When the muscles relax, the lactic acid is removed and the pain goes away.*

4–2 Section Review

1. What is respiration? How does it enable organisms to use the energy in food?
2. Why are the mitochondria appropriately called the powerhouses of the cell?
3. How are aerobic and anaerobic respiration the same? How are they different?

Connection—*You and Your World*
4. Brewers are careful to keep oxygen away from yeast cells during the process of making beer or wine. Based on this information, what conclusions can you draw about the type of respiration yeasts undergo when oxygen is present?

Life in the Darkness

Although we have sent satellites to explore the far reaches of our galaxy, there are places on Earth that still have not been fully explored—not yet, anyway. For example, our knowledge of ocean floor *ecosystems* in the deepest parts of the oceans is still limited.

One tool scientists can use to explore the ocean floor is a deep-sea submersible such as the *Alvin*. Deep-sea submersibles can carry scientists to the deepest parts of the ocean. Not long ago scientists studying the Pacific Ocean floor off the coast of the Galapagos Islands near South America made an astonishing discovery. At a depth of 2700 meters they found that the water, which should have been near-freezing, was quite hot. The heat was generated by volcanic vents, or holes, in the ocean floor out of which superheated water erupted.

Hot water may not seem that exciting, but what the scientists found near the vents certainly was. Much to their surprise, unusual organisms such as tube worms, huge clams, and eyeless crabs were quite at home in this environment.

After studying the organisms, scientists realized that there were bacteria living within the bodies of some of the organisms, particularly in the tube worms. These bacteria were able to absorb chemicals, mainly hydrogen sulfide, carried up from under the Earth's crust by the gushing water in the vents. The bacteria could then use the energy in the atoms of hydrogen sulfide to produce food molecules such as glucose. That is, the bacteria used hydrogen sulfide instead of sunlight as their source of energy. The other organisms living nearby the vents then used the food made by the bacteria.

Today scientists call such bacteria *chemosynthetic* bacteria. Chemosynthetic bacteria, as the name implies, use the energy in certain chemicals to synthesize (to make) food. They are among the few organisms on Earth that do not rely on sunlight and photosynthesis for survival.

Laboratory Investigation

Comparing Photosynthesis and Respiration

Problem

What is the relationship between photosynthesis and respiration?

Materials *(per group)*

100-mL graduated cylinder
bromthymol blue solution
2 125-mL flasks
straw
2 *Elodea*
2 #5 rubber stoppers
light source

Procedure 🝠 🖾

1. Using the graduated cylinder, pour 100 mL of bromthymol blue solution into each flask. **CAUTION:** *Bromthymol blue is a dye and can stain your hands and clothing.*

2. Insert one end of the straw into the bromthymol blue solution in one of the flasks. Gently blow through the straw. Keep blowing gently until there is a change in the color of the solution. (Bromthymol blue turns yellow in the presence of carbon dioxide.) Repeat this procedure with the other flask.

3. Place a sprig of *Elodea* into each flask. *Elodea* is an organism that will perform photosynthesis when placed in sunlight. Put a stopper in each of the flasks.

4. Place one flask in the dark for 24 hours and the other flask on a sunny windowsill for the same amount of time.

Observations

1. What was the color of the bromthymol blue solution before and after you exhaled into it?

2. What was the color of the bromthymol blue solution in the flask placed in the dark for 24 hours? In the flask on the windowsill?

Analysis and Conclusions

1. How was the carbon dioxide you exhaled into the bromthymol blue solution produced in your body?

2. Why was *Elodea* placed in both flasks?

3. How can you explain your observations for each flask?

4. How are photosynthesis and respiration related?

5. **On Your Own** Design an experiment to see if temperature changes have any effect on the results of this experiment. With your teacher's permission, carry out the experiment.

Summarizing Key Concepts

4–1 Photosynthesis: Capturing and Converting Sunlight

▲ An organism needs food for energy, to repair cell parts, and to make new cell parts.

▲ The energy in food is locked in the bonds that link atoms together.

▲ Photosynthesis is the process in which the energy of sunlight is used to produce food.

▲ The first step in photosynthesis occurs when energy in the form of sunlight is captured by the chlorophyll in the chloroplasts.

▲ The second step in photosynthesis is the process in which the radiant energy of sunlight is converted into the chemical energy found in food.

▲ The basic food produced by photosynthesis is the simple sugar called glucose. Using glucose and other substances, an organism can produce all the chemicals of life.

▲ In addition to sunlight, green plants need carbon dioxide and water to perform photosynthesis.

▲ Oxygen is a waste product of photosynthesis.

▲ Oxygen produced during photosynthesis is required by nearly all living things.

▲ Organisms that can make their own food are called autotrophs. Organisms that cannot make their own food and rely on other sources of food for energy are called heterotrophs.

4–2 Respiration: Using the Energy in Food

▲ Respiration is the process in which simple food substances such as glucose are broken down and the energy they contain is released.

▲ Aerobic respiration, which is the process most organisms undergo, occurs in the mitochondria and requires oxygen.

▲ During the process of respiration, carbon dioxide and water are given off as waste products.

▲ In many respects, the process of respiration is the opposite of the process of photosynthesis.

▲ Some organisms utilize fermentation (anaerobic respiration), which is a process in which food is broken down and energy released without need of oxygen.

Reviewing Key Terms

Define each term in a complete sentence.

4–1 Photosynthesis: Capturing and Converting Sunlight
photosynthesis
stomata
autotroph
heterotroph

4–2 Respiration: Using the Energy in Food
respiration
ATP
fermentation

Chapter Review

Content Review

Multiple Choice

Choose the letter of the answer that best completes each statement.

1. The energy in food is locked up within its
   a. nutrients.
   c. chloroplasts.
   b. chemical bonds.
   d. mitochondria.
2. The green pigment found in plants is called
   a. chloroplast.
   b. photosynthetic pigment.
   c. chlorophyll.
   d. organelle.
3. Two substances produced during photosynthesis are
   a. glucose and oxygen.
   b. glucose and carbon dioxide.
   c. glucose and water.
   d. oxygen and water.
4. Two substances required for photosynthesis are
   a. carbon dioxide and oxygen.
   b. water and oxygen.
   c. carbon dioxide and water.
   d. water and glucose.

5. An organism that cannot make its own food is called a(an)
   a. green plant.
   b. heterotroph.
   c. autotroph.
   d. nonphotosynthetic organism.
6. The basic food produced by photosynthesis is a
   a. sugar.
   c. protein.
   b. starch.
   d. nucleic acid.
7. The process in which food is broken down and the energy in the food is released is called
   a. photosynthesis.
   c. respiration.
   b. ATP.
   d. digestion.
8. The end products of respiration include
   a. water and ATP.
   b. glucose and oxygen.
   c. glucose and ATP.
   d. oxygen and ATP.

True or False

If the statement is true, write "true." If it is false, change the underlined word or words to make the statement true.

1. The <u>sun</u> is the primary source of energy for almost all living things.
2. <u>All the steps</u> in photosynthesis occur in sunlight.
3. Green plants perform photosynthesis in order to obtain <u>oxygen</u>.
4. Yeasts perform a type of respiration called <u>fermentation</u>.
5. The pigment that makes green plants appear green is called <u>chloroplast</u>.
6. Plants require <u>water</u> and <u>oxygen</u> to perform photosynthesis.
7. <u>Autotrophs</u> are organisms that can make their own food.

Concept Mapping

Complete the following concept map for Section 4–1. Refer to pages D6–D7 to construct a concept map for the entire chapter.

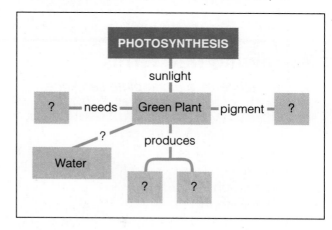

Concept Mastery

Discuss each of the following in a brief paragraph.

1. In order, describe the basic steps of photosynthesis.
2. Name ten living things you see every day. Identify each organism on your list as either an autotroph or a heterotroph.
3. How are photosynthesis and respiration related?
4. Explain why you would find more mitochondria in a muscle cell than in a skin cell.
5. Some people interchange the terms breathing and respiration. Explain why these are not the same in terms of a cell. What relationship does breathing have to cellular respiration?
6. Discuss the importance of the waste product oxygen that is produced during photosynthesis.

Critical Thinking and Problem Solving

Use the skills you have developed in this chapter to answer each of the following.

1. **Assessing concepts** Which is better, respiration or fermentation? Explain your answer.
2. **Drawing conclusions** Some scientists feel that dinosaurs became extinct when a cloud of dust and gas kept sunlight from penetrating the atmosphere for a long period of time. Based on what you now know about photosynthesis, is this a reasonable conclusion? If so, explain how a lack of sunlight might have affected the dinosaurs.

3. **Making inferences** Scientists often refer to the stages of photosynthesis as the light and dark reactions. Explain why these terms are appropriate.
4. **Making an outline** Outline the processes of photosynthesis and respiration.
5. **Making comparisons** Compare the general equations for photosynthesis and respiration. Why might the two processes be considered a cycle?
6. **Using the writing process** Imagine that humans were green and could perform photosynthesis. Design an advertisement for a travel agency that encourages people to take a Caribbean cruise. Your advertisement can be written or in the form of a poster.

GAZETTE

FROM BIRD SONGS TO REBORN BRAINS

As he roamed his family's ranch in Argentina, listening to the songs of birds, the young Fernando Nottebohm felt the first stirrings of becoming a naturalist. The birds' songs intrigued him. As he walked, he would play the game of matching each song he heard to the type of bird that produced it. In this way, he created his own private hit parade—his own "Top Twenty." Any unfamiliar melody would arouse his curiosity even more, so that he would follow the sound until he discovered its source.

"That's how a naturalist develops, I think," says Nottebohm. "You just have this joy in poking around and spying on things, always hoping to find something novel." And many years later, that's exactly what Nottebohm is doing—poking around and discovering things. Only, it is birds' brains (not songs) that the naturalist-turned-zoologist is now researching. And you can be sure that plenty of novel situations have presented themselves over the course of this researcher's career.

Working at Rockefeller University in New York and directing the school's Field Research Center for Ecology and Ethology, Nottebohm has extended his bird studies into an area that is both intriguing and astonishing. In addition, his discoveries hold promise for applications which are nothing less than revolutionary.

The area of Nottebohm's study is neurogenesis, which is the birth of new neurons. Neurons are nerve cells in the brain. As far as anyone knows, nerve cells in the human brain that have been destroyed by injury or disease do not regenerate (grow back). Functions once carried out by these neurons can often be performed by other, healthy cells in the brain, but the dead cells themselves can never be replaced by new ones. All the neurons you will ever have were formed by the time you were six months old. But according to Nottebohm's research, an adult bird can generate as many as 20,000 new neurons in a single day! If Nottebohm and his fellow researchers can figure out exactly what factors control neurogenesis in birds, they may be on their way to figuring out how to trigger this same process in humans.

Nottebohm's journey into the mysteries of the human brain has taken him far afield of his original ambitions. "I was mainly just interested in bird behavior and evolution at the beginning," he says. That interest in evolution was sparked at age 17, when Nottebohm first read Charles Darwin's *Voyage of the Beagle*. The content, vibrancy, and romanticism that Darwin brought to zoology convinced Nottebohm to try to make a career out of his love for nature. But the prospects for earning a living as a bird zoologist in Argentina were extremely poor. So with the encouragement of his father, Nottebohm decided to study agriculture, run the family ranch, and pursue his bird interests on the weekends.

After studying for a year at the University of Nebraska, Nottebohm transferred to the University of California at Berkeley. There he became captivated by the zoology lectures of Peter Marler, who would become internationally famous for his work on vocal communication in songbirds, monkeys, and apes. It was not long before Nottebohm, as a graduate student working with Marler, returned to his original fascination—bird songs. Only this time his approach was different: His research examined how a membrane in the bird's voice box, or syrinx, vibrates as the animal sings.

> ▼ **The tracings show the songs of two canaries recorded with a sound spectrograph. One canary suffered damage to the right side of its brain, which did not greatly affect its song (top). The other had damage to the left side of its brain, after which its song became distorted (bottom).**

As is often the case in science, the search for an answer to one particular question generates new questions and new areas of research. And so Nottebohm soon found himself investigating more than just one membrane of a bird's syrinx. The link between the brain and the syrinx became the target of his research. Identifying the song-control areas of the brain and tracing the neural pathways between brain and syrinx was a project that took Nottebohm five years.

But the series of discoveries was not to end there. Along the way, Nottebohm and neurobiologist Arthur Arnold discovered that in canaries male and female brains are different. And in male canaries the size of the song-control areas changes dramatically with the seasons—increasing in size in the spring and shrinking at the end of the summer. Then finally, the most amazing discovery of all: Neurogenesis occurs in adult birds.

Today, a major aim of Nottebohm and his research team is to identify the genes that govern the birth of new neurons in bird brains and to determine the chemical signals that turn these genes on. As Nottebohm sees it, because evolution tends to be conservative, these genes should also be present in our own brains. Indeed, it is probably just these genes that control neurogenesis during the early development of humans. "The challenge," says Nottebohm, "is to figure out how to reawaken this genetic potential that may be lying dormant within certain brain cells in mammals, including ourselves."

BIRD 54

BIRD 97

HUMAN-GROWTH HORMONE:
USE OR ABUSE?

Soon after dinner six days a week, Marco Oriti tries to relax as his mother injects him with a synthetic hormone. The injections are not to cure a disease; Marco is actually quite healthy. He just wants to be taller.

The hormone Marco receives is called human-growth hormone, or HGH. It is HGH that controls how tall each of us will become. HGH is normally produced in the body, particularly during the teenage years when most young people have a growth spurt. Synthetic HGH is almost identical to natural HGH, except that it is produced in the laboratory and not in the body.

BACTERIA: HGH PRODUCTION FACTORIES

The production of synthetic HGH is due to recent advances in biotechnology. In a procedure that is becoming more and more commonplace, the gene that controls the production of HGH is inserted into bacteria. In this way the bacteria can be made to produce human-growth hormone. (It's still called human even though it is made in a bacterial cell.) As the bacteria multiply, each succeeding generation contains the gene for HGH production. In time a colony of HGH-producing bacteria forms. In a sense the colony is an HGH production line.

Using biotechnology, HGH can now be obtained in plentiful quantities. And although the cost is still high—Marco's injections cost about $15,000 a year— many people feel the money is well spent.

Growth Plates

▲ **Growth plates, which are located at the ends of the long bones of the body, are the sites at which bones grow in length.**

HGH DEFICIENCIES

The development of synthetic HGH was intended primarily for those children who do not produce adequate amounts of the hormone. Such children have an HGH deficiency. Most people agree that children with an HGH deficiency should receive the new drug. The number of children who fit into this category, however, is quite limited. Most children do produce enough HGH. Will all these children grow tall? No, they will grow to the height that is normal for them. People vary in size, just as they do in skin color, eye color, and type of hair. Differences are a part of being human.

WHO SHOULD RECEIVE HGH?

Marco Oriti is a normal child. He does not have an HGH deficiency. He simply wants to grow taller. And that is the core of the issue facing doctors today. Many doctors believe that HGH should not be given to healthy children who merely want to change their normal height. These doctors point out that being tall or short is not a disease and should not be treated as one. They also note that the use of the drug is still experimental and that side effects may occur. (Marco already has one side effect—he often loses his appetite.) In addition, the cost of the drug is difficult for most families to afford. Marco's treatment over the years will cost about $150,000.

Many of these doctors assert that the use of HGH is being promoted simply to make money for the pharmaceutical companies that manufacture the drug. After all, since few children have an HGH deficiency, the production of the drug would not be economically feasible unless a larger market could be found. And that larger market, say the doctors, are normal children who are being lured by the promise of being taller.

Other doctors argue that HGH exists, and if people want to use it they should be free to do so. They point out that whether we like it or not, height is a significant factor in our lives. Statistics bear these doctors out.

▲ During the first 15 or 16 years of life, children grow at different rates. Body growth is usually complete in the late teens, at which time the production of growth hormone drops.
▽ Too little growth hormone results in a condition called pituitary dwarfism.

▲ Soldiers grouped according to differences in height form a living graph. The soldiers in the middle of the group represent the average height of the group.

It has been shown that job candidates have a better chance of getting a job if they are tall; politicians are more likely to be elected if they are tall; even starting salaries are higher for taller people.

People against the use of HGH contend, however, that instead of changing our bodies with experimental drugs, we should be changing the way we view ourselves. That is, we should learn to accept what is normal for each of us. More than accept—we should delight in the individual differences that distinguish one human from another.

What is your opinion? Should drugs be used to change people from what they are going to be into what they want to be, no matter the risks? Or should we learn to love ourselves for what we are—and not for what we would like to be?

▼ The pituitary gland is a bean-sized structure at the base of the skull. Growth hormone is secreted by the anterior pituitary.

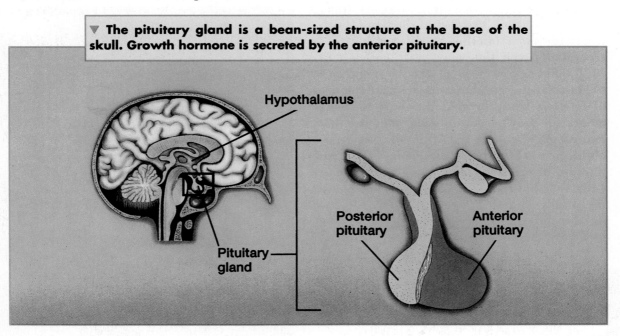

Hypothalamus

Pituitary gland

Posterior pituitary

Anterior pituitary

THE GREEN PEOPLE OF
SOLARON

The temperature outside the bubble is over 100°C. The 30-hour day on the planet Solaron is only half over.

Solaron, which is in a galaxy 20 light-years from Earth, was colonized by 100 humans in the year 2186. Now, in the year 2215, the colony has a population of 2000.

These people can live here because the colony is covered by a huge plastic bubble. Inside the bubble are schools, stores, offices and other buildings. The artificial atmosphere is made up of oxygen, carbon dioxide, nitrogen, and the other gases that make up Earth's atmosphere. Outside the bubble, there is no air. Except for a single bright sun that always shines, the sky is black. The

hot surface of the planet is covered with red sand, rocks, and boulders.

Zara Starr looks out the window of her ninth-grade classroom. Through the window, she can see the clear plastic bubble that encloses her world. She can see the horizon far off in the distance, where the black sky meets the red soil.

Suddenly, the bell rings, and Zara and her twin brother Lars gather their books and walk quickly to their lockers. As they pass through the corridor, some new students, who have just arrived at the colony, gape at the Starr twins. The new students have never seen green people before. The twins ignore the stares of the new students. They don't mind being different. In fact, Zara and

Lars are proud of being green. They are part of an important experiment that appears to be successful.

GREEN IS GOOD

Scientists on Earth wanted to know whether human skin cells could act like tiny plants by using the energy of sunlight to make glucose from carbon dioxide and water. This process called photosynthesis produces glucose, a sugar plants use as food. If human cells could make their own food, scientists thought, colonists on planets such as Solaron would not have to grow so much food. With less need for grown food, the colonists on Solaron could have smaller farms. This would save precious space under the colony's bubble and conserve limited energy resources. Green people could help Solaron in another way, too. Oxygen is a byproduct of photosynthesis. So human photosynthesis would be a source of oxygen. Green people would give off oxygen—just as green plants do—for other people to breathe.

To begin this great experiment, scientists first made copies of genes that control the tiny green food-making organs in plants. The techniques for gene copying, called genetic engineering, had been developed in the 1980s. At that time, scientists put the copies of these plant genes into special viruses. These viruses were similar to the viruses that cause the common cold. But researchers had changed the viruses slightly so that they would go only to skin cells after being injected into human volunteers. When the viruses reached the skin cells, the plant genes inside the viruses joined the human genes inside the skin cells. As the plant genes began to work inside the skin cells, the volunteers slowly turned green. That was because the plant genes were making chlorophyll. Chlorophyll is the special green chemical that captures the energy of sunlight to use for photosynthesis. Because the sun shines all the time on Solaron, the green skin of the volunteers carries on photosynthesis every day.

Green people shed old skin cells, just as other people do. And just like other people's, their bodies make new skin cells. Each new cell is an exact copy of the parent cell. So green people never lose their color.

Because Zara, Lars, and their parents are green, they live in a special house with a clear plastic roof. Sunlight pours down through this roof and onto the members of the Starr family as they go about their chores.

When the Starrs wake up in the morning, they are never hungry because their skin cells have made glucose during the night. Some of the glucose leaves the skin cells and is stored in the liver. When the body needs food during the day, the liver releases some of the stored glucose. So Zara and Lars never have to eat foods rich in sugar.

At lunch time, when their schoolmates are in the cafeteria, the Starr twins and other green students go to a special room called the solar room. It has a clear plastic ceiling that allows the students to get plenty of sunlight. The students read, talk, or just lie back and close their eyes, imagining they are relaxing on a beach on Earth. Meanwhile, their skin cells are storing the sun's energy.

THE NEW KIDS IN TOWN

Most students are used to seeing green students sitting next to them in class. But today, all the students are staring at some new volunteers for the photosynthesis experiment. These new students have red skin. Scientists succeeded in putting a special pigment, a colored chemical called anthocyanin, into skin cells, along with chlorophyll. In the leaves of plants and trees, anthocyanin pigments are different colors under different conditions. In the fall, when the green chlorophyll of leaves breaks down, the red,

blues, and purples of anthocyanins start to show. The leaves are said to "change color."

The new students have a red pigment in their skin cells that hides the green color of the chlorophyll. However, the chlorophyll still works as well as it does in green people. Scientists want to experiment with different skin colors. That way, volunteers could have a choice of what color they want to be. The scientists hope that being able to choose from among many colors will make more people volunteer for the experiment. Then there will be more people on Solaron making their own food and producing oxygen.

In a few days, no one will notice that the red people are different from anyone else. Everybody is too busy with schoolwork, dances, parties, sports, and families. After all, says one of the students, "It doesn't matter whether we're green or red. We're all just human."

For Further Reading

> If you have been intrigued by the concepts examined in this textbook, you may also be interested in the ways fellow thinkers—novelists, poets, essayists, as well as scientists—have imaginatively explored the same ideas.

Chapter 1: The Nature of Life

L'Engle, Madeleine. *A Ring of Endless Light*. New York: Farrar, Straus & Giroux.

Pirotta, Saviour. *The Flower from Outer Space*. New York: P. Bedrick Books.

Rifas, Leonard. *Food First Comic*. San Francisco: Institute for Food and Development Policy.

Ryder, Donald G. *The Inside Story: Living and Learning Through Life's Storms*. Pleasant Hill, CA: Ryder.

Watson, James D. *The DNA Story*. San Francisco: W.H. Freeman.

Chapter 2: Cell Structure and Function

Cronin, A. J. *The Citadel*. Boston: Little, Brown.

Davis, Natalie. *The Space Twin*. Nashville, TN: Winston-Derek.

Lasky, Kathryn. *The Bone Wars*. New York: Morrow Junior Books.

Rowe, H. Edward. *Microscopic Monster: The Tricky, Devastating AIDS Virus*. Culpepper, VA: National AIDS Prevention Institute.

Chapter 3: Cell Processes

Cone, Molly. *Dance Around the Fire*. Boston: Houghton Mifflin.

Gonick, Larry and Mark Wheelis. *Cartoon Guide to Genetics*. Fort Ann, NY: Barnes & Noble Books.

Pascal, Francine. *Tug of War*. New York: Bantam Books.

Whelan, Gloria. *The Pathless Woods*. Philadelphia: Lippincott.

Chapter 4: Cell Energy

Hackman, Martha. *The Lost Forest*. San Marcos, CA: Green Tiger Press.

Llewellyn, Richard. *How Green Was My Valley*. New York: Dell.

Place, Marion T. *Mount St. Helens*. New York: Putnam Publishing Group.

Rose, Walter. *The Village Carpenter*. New York: New Amsterdam Books.

Wrightson, Patricia. *The Nargun and the Stars*. New York: Macmillan.

Activity Bank

Welcome to the Activity Bank! This is an exciting and enjoyable part of your science textbook. By using the Activity Bank you will have the chance to make a variety of interesting and different observations about science. The best thing about the Activity Bank is that you and your classmates will become the detectives, and as with any investigation you will have to sort through information to find the truth. There will be many twists and turns along the way, some surprises and disappointments too. So always remember to keep an open mind, ask lots of questions, and have fun learning about science.

HYDRA DOING?

Hydras are members of the phylum *Cnidaria*. Like all cnidarians, hydras have soft bodies made up of two layers of cells. They also have stinging tentacles arranged in circles around their mouth. Do hydras have the characteristics that all forms of life share? To answer this question, you will need a medicine dropper, a culture of hydras, a depression slide, a microscope, a toothpick, and some fish food.

Procedure 🧪🐀

1. Using a medicine dropper, remove a drop of the hydra culture from the bottom of the culture jar.
2. Place the drop on a depression slide.

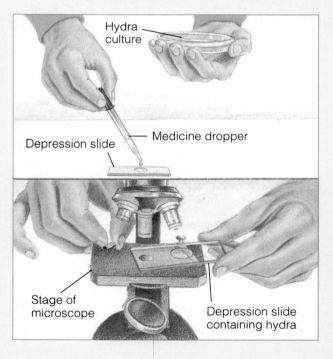

3. Put the slide on the stage of a microscope. Using the low-power objective, locate a hydra. Draw and label what you observe.

4. Carefully touch the hydra's tentacles with a toothpick. Observe what happens.

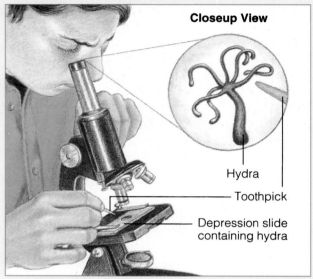

5. Place a tiny amount of fish food near the hydra and see what happens.

Observations

1. What happened to the hydra when you touched it with the toothpick?
2. How does the hydra move?
3. Describe how the hydra eats.

Analysis and Conclusions

1. What characteristics of living things does the hydra exhibit?
2. Does the hydra react as a whole organism or does just part of the hydra react? Explain.

Going Further

Place a drop of vinegar (weak acid) near a hydra. Observe what happens.

NOW YOU SEE IT—NOW YOU DON'T

Cells are the basic units of structure and function of all living things. Cells contain certain structures that carry out life processes. One such structure is the cell membrane. The cell membrane regulates what goes into and what comes out of a cell. In plants, the cell membrane is just inside the cell wall. For this reason, it is sometimes difficult to see the cell membrane. To help you see the cell membrane in plants a little better, why not try this activity.

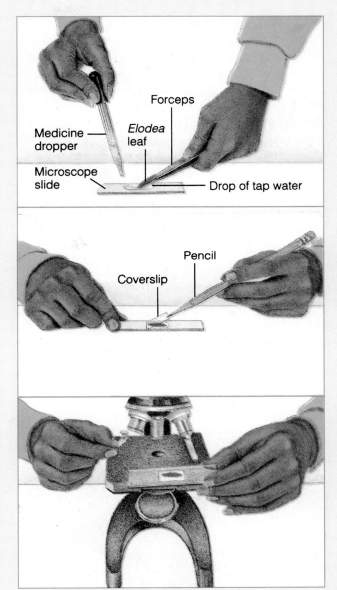

Materials

2 medicine droppers	pencil
microscope slide	coverslip
forceps	microscope
Elodea leaf	salt solution
	paper towel

Procedure 🧪

1. Using a medicine dropper, place one drop of tap water on a microscope slide.

2. With forceps, place an *Elodea* leaf in the drop of water. Cover with a coverslip.

3. Observe the leaf under both low and high powers of the microscope. Note the location of the chloroplasts in relation to the cell wall. Draw a diagram of a plant cell, labeling the structures you observe.

(continued)

4. Using the other medicine dropper, add a drop of the salt solution along one edge of the coverslip. Place a piece of

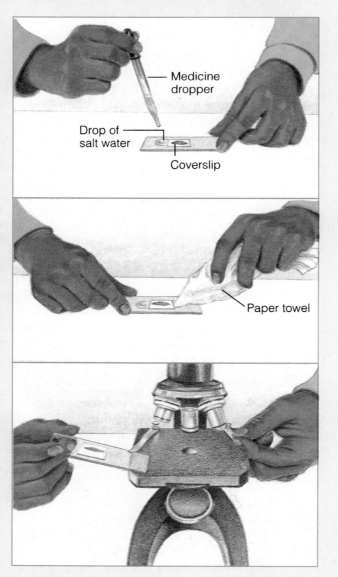

Medicine dropper

Drop of salt water

Coverslip

Paper towel

paper towel along the opposite edge of the coverslip as shown in the center diagram on this page. The tap water will soak into the paper towel, drawing the salt solution under the coverslip.

5. Observe the leaf under both low and high powers of the microscope. Again note the location of the chloroplasts in relation to the cell wall. Draw another diagram of the cell, labeling the cell wall, cell membrane, and chloroplasts.

6. Repeat step 4 using a drop of tap water instead of a drop of the salt solution. Observe the appearance of the cells.

Observations

1. Where were the chloroplasts located in the cell when tap water was added?

2. Where were the chloroplasts located when the salt solution was added?

Analysis and Conclusions

1. In which direction did water move when the salt solution was added?

2. What happened to the cell when the salt solution was added?

3. What happened to the cell when tap water was added for the second time?

Going Further

Repeat this activity using a sugar (glucose) solution. Does a glucose solution have the same effect on a cell as a salt solution does?

One of the most important processes that occurs in a cell is diffusion. Diffusion regulates what enters and leaves the cell. To see how diffusion actually occurs, try this activity.

Materials

50 mL household ammonia
wide-mouthed jar
piece of cheese-cloth (15 cm × 15 cm)

rubber band
spatula
gelatin "cell"
clock with second indicator

Procedure 🧪 📷 👁

1. Select one member of the group to act as a Principal Investigator, a second member to act as a Timer, and a third member to act as a Recorder. The remaining members of the group will be the Observers. Be sure you understand your role in the activity before you continue.

2. Carefully pour 50 mL of ammonia in a wide-mouthed jar. **CAUTION:** *Keep the ammonia away from your skin and do not inhale its vapors. Keep the room well ventilated.*

3. Place the cheesecloth over the mouth of the jar. Hold it in place by putting a rubber band around the jar and the cheesecloth.

4. Using a spatula, place a gelatin "cell" on top of the cheesecloth as shown in the diagram. The gelatin "cell" contains a chemical called phenolphthalein. Phenolphthalein indicates the presence of bases such as ammonia. If a base is present, phenolphthalein will turn pink. **Note:** *Do not allow any ammonia to come into direct contact with the gelatin "cell."*

Household ammonia

Wide-mouthed jar

Cheesecloth

Rubber band

50 mL of household ammonia

Gelatin "cell"

(continued)

5. Note the color of the gelatin "cell" immediately after placing it on the cheesecloth. Record its color in a data table similar to the one shown.

6. Observe the gelatin "cell" every 2 minutes for a total of 10 minutes. Record your observations in your data table.

Observations

DATA TABLE

Time (min)	Color Change
0	
2	
4	
6	
8	
10	

Analysis and Conclusions

1. Does the ammonia diffuse into the gelatin "cell" or does the material in the gelatin "cell" diffuse into the ammonia in the jar? How do you know?

2. What causes the gelatin "cell" to undergo changes without coming into direct contact with the ammonia?

3. Compare your results with those of your classmates. Are they similar? Different? If they are different, explain why.

In order to carry out photosynthesis, a green plant must have light. But how much light? To show how different intensities (amounts) of light affect photosynthesis, why not do this activity.

Materials

test tube
400-mL beaker
freshly cut *Elodea* sprig
forceps
bright light

sodium bicarbonate solution
hand lens
clock with second indicator

Procedure 🧪 📷

1. Select one member of the group for each of the following roles: Principal Investigator, Timer, Counter, and Observer/Recorder. Make sure you understand your role in the activity before you continue.

2. Completely fill a test tube and a beaker with a sodium bicarbonate solution. Sodium bicarbonate will provide a source of carbon dioxide.

3. Using forceps, place an *Elodea* sprig about halfway down in the test tube. Be sure that the cut end of the plant points downward in the test tube.

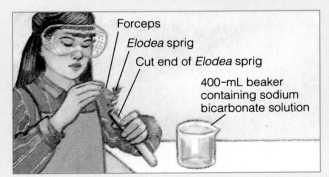

Forceps
Elodea sprig
Cut end of *Elodea* sprig
400-mL beaker containing sodium bicarbonate solution

4. Cover the mouth of the test tube with your thumb and turn the test tube upside down. Try not to trap any air bubbles in the test tube.

5. Place the mouth of the test tube under the surface of the sodium bicarbonate solution in the beaker. Remove your thumb from the mouth of the test tube.

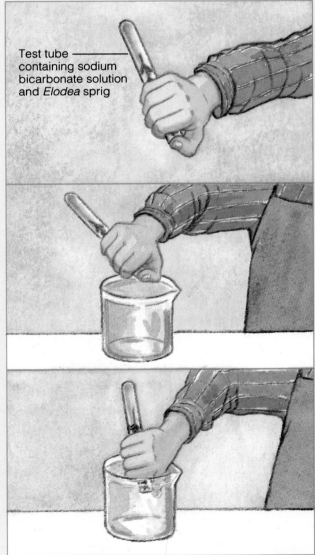

Test tube containing sodium bicarbonate solution and *Elodea* sprig

(continued)

6. Gently lower the test tube inside the beaker so that the test tube leans against the side of the beaker.

7. Put the beaker in a place where it will receive normal room light. Using a hand lens, count the number of bubbles produced by the *Elodea* in the test tube for 5 minutes. Record the number of bubbles in a data table similar to the one shown.

8. Turn down the lights in the room and count the number of bubbles again for 5 minutes. Record the number in your data table.

9. Turn up the lights in the room and shine a bright light on the *Elodea*. Count the number of bubbles produced in 5 minutes. Record the number in your data table.

Observations

DATA TABLE

Light Intensity	Number of Bubbles Produced in 5 Minutes
Room light	
Dim light	
Bright light	

Analysis and Conclusions

1. From what part of the *Elodea* did the bubbles emerge?
2. How does counting bubbles measure the rate of photosynthesis?
3. When was the greatest number of bubbles produced? The least?
4. How does the intensity of light affect the rate of photosynthesis?
5. How do your results compare with those of your classmates? Are they similar? Different? How can you account for any differences in the numbers of bubbles produced? Can you identify any trends even if the actual numbers differ?

Going Further

Perform the activity again using different colors of light. What effect does each color have on the rate of photosynthesis?

The metric system of measurement is used by scientists throughout the world. It is based on units of ten. Each unit is ten times larger or ten times smaller than the next unit. The most commonly used units of the metric system are given below. After you have finished reading about the metric system, try to put it to use. How tall are you in metrics? What is your mass? What is your normal body temperature in degrees Celsius?

Commonly Used Metric Units

Length   The distance from one point to another

meter (m)	A meter is slightly longer than a yard.
	1 meter = 1000 millimeters (mm)
	1 meter = 100 centimeters (cm)
	1000 meters = 1 kilometer (km)

Volume   The amount of space an object takes up

liter (L)	A liter is slightly more than a quart.
	1 liter = 1000 milliliters (mL)

Mass   The amount of matter in an object

gram (g)	A gram has a mass equal to about one paper clip.
	1000 grams = 1 kilogram (kg)

Temperature   The measure of hotness or coldness

degrees	0°C = freezing point of water
Celsius (°C)	100°C = boiling point of water

Metric–English Equivalents

2.54 centimeters (cm) = 1 inch (in.)
1 meter (m) = 39.37 inches (in.)
1 kilometer (km) = 0.62 miles (mi)
1 liter (L) = 1.06 quarts (qt)
250 milliliters (mL) = 1 cup (c)
1 kilogram (kg) = 2.2 pounds (lb)
28.3 grams (g) = 1 ounce (oz)
$°C = 5/9 \times (°F - 32)$

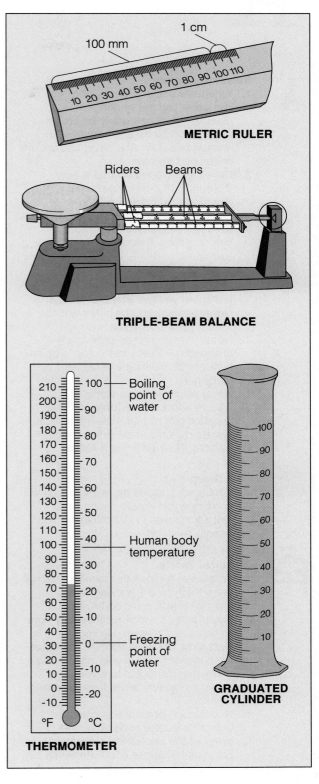

METRIC RULER

Riders   Beams

TRIPLE-BEAM BALANCE

Boiling point of water

Human body temperature

Freezing point of water

°F   °C

THERMOMETER

GRADUATED CYLINDER

Glassware Safety

1. Whenever you see this symbol, you will know that you are working with glassware that can easily be broken. Take particular care to handle such glassware safely. And never use broken or chipped glassware.
2. Never heat glassware that is not thoroughly dry. Never pick up any glassware unless you are sure it is not hot. If it is hot, use heat-resistant gloves.
3. Always clean glassware thoroughly before putting it away.

Fire Safety

1. Whenever you see this symbol, you will know that you are working with fire. Never use any source of fire without wearing safety goggles.
2. Never heat anything—particularly chemicals—unless instructed to do so.
3. Never heat anything in a closed container.
4. Never reach across a flame.
5. Always use a clamp, tongs, or heat-resistant gloves to handle hot objects.
6. Always maintain a clean work area, particularly when using a flame.

Heat Safety

Whenever you see this symbol, you will know that you should put on heat-resistant gloves to avoid burning your hands.

Chemical Safety

1. Whenever you see this symbol, you will know that you are working with chemicals that could be hazardous.
2. Never smell any chemical directly from its container. Always use your hand to waft some of the odors from the top of the container toward your nose—and only when instructed to do so.
3. Never mix chemicals unless instructed to do so.
4. Never touch or taste any chemical unless instructed to do so.
5. Keep all lids closed when chemicals are not in use. Dispose of all chemicals as instructed by your teacher.

6. Immediately rinse with water any chemicals, particularly acids, that get on your skin and clothes. Then notify your teacher.

Eye and Face Safety

1. Whenever you see this symbol, you will know that you are performing an experiment in which you must take precautions to protect your eyes and face by wearing safety goggles.
2. When you are heating a test tube or bottle, always point it away from you and others. Chemicals can splash or boil out of a heated test tube.

Sharp Instrument Safety

1. Whenever you see this symbol, you will know that you are working with a sharp instrument.
2. Always use single-edged razors; double-edged razors are too dangerous.
3. Handle any sharp instrument with extreme care. Never cut any material toward you; always cut away from you.
4. Immediately notify your teacher if your skin is cut.

Electrical Safety

1. Whenever you see this symbol, you will know that you are using electricity in the laboratory.
2. Never use long extension cords to plug in any electrical device. Do not plug too many appliances into one socket or you may overload the socket and cause a fire.
3. Never touch an electrical appliance or outlet with wet hands.

Animal Safety

1. Whenever you see this symbol, you will know that you are working with live animals.
2. Do not cause pain, discomfort, or injury to an animal.
3. Follow your teacher's directions when handling animals. Wash your hands thoroughly after handling animals or their cages.

One of the first things a scientist learns is that working in the laboratory can be an exciting experience. But the laboratory can also be quite dangerous if proper safety rules are not followed at all times. To prepare yourself for a safe year in the laboratory, read over the following safety rules. Then read them a second time. Make sure you understand each rule. If you do not, ask your teacher to explain any rules you are unsure of.

Dress Code

1. Many materials in the laboratory can cause eye injury. To protect yourself from possible injury, wear safety goggles whenever you are working with chemicals, burners, or any substance that might get into your eyes. Never wear contact lenses in the laboratory.

2. Wear a laboratory apron or coat whenever you are working with chemicals or heated substances.

3. Tie back long hair to keep it away from any chemicals, burners and candles, or other laboratory equipment.

4. Remove or tie back any article of clothing or jewelry that can hang down and touch chemicals and flames.

General Safety Rules

5. Read all directions for an experiment several times. Follow the directions exactly as they are written. If you are in doubt about any part of the experiment, ask your teacher for assistance.

6. Never perform activities that are not authorized by your teacher. Obtain permission before "experimenting" on your own.

7. Never handle any equipment unless you have specific permission.

8. Take extreme care not to spill any material in the laboratory. If a spill occurs, immediately ask your teacher about the proper cleanup procedure. Never simply pour chemicals or other substances into the sink or trash container.

9. Never eat in the laboratory.

10. Wash your hands before and after each experiment.

First Aid

11. Immediately report all accidents, no matter how minor, to your teacher.

12. Learn what to do in case of specific accidents, such as getting acid in your eyes or on your skin. (Rinse acids from your body with lots of water.)

13. Become aware of the location of the first-aid kit. But your teacher should administer any required first aid due to injury. Or your teacher may send you to the school nurse or call a physician.

14. Know where and how to report an accident or fire. Find out the location of the fire extinguisher, phone, and fire alarm. Keep a list of important phone numbers—such as the fire department and the school nurse—near the phone. Immediately report any fires to your teacher.

Heating and Fire Safety

15. Again, never use a heat source, such as a candle or burner, without wearing safety goggles.

16. Never heat a chemical you are not instructed to heat. A chemical that is harmless when cool may be dangerous when heated.

17. Maintain a clean work area and keep all materials away from flames.

18. Never reach across a flame.

19. Make sure you know how to light a Bunsen burner. (Your teacher will demonstrate the proper procedure for lighting a burner.) If the flame leaps out of a burner toward you, immediately turn off the gas. Do not touch the burner. It may be hot. And never leave a lighted burner unattended!

20. When heating a test tube or bottle, always point it away from you and others. Chemicals can splash or boil out of a heated test tube.

21. Never heat a liquid in a closed container. The expanding gases produced may blow the container apart, injuring you or others.

22. Before picking up a container that has been heated, first hold the back of your hand near it. If you can feel the heat on the back of your hand, the container may be too hot to handle. Use a clamp or tongs when handling hot containers.

Using Chemicals Safely

23. Never mix chemicals for the "fun of it." You might produce a dangerous, possibly explosive substance.

24. Never touch, taste, or smell a chemical unless you are instructed by your teacher to do so. Many chemicals are poisonous. If you are instructed to note the fumes in an experiment, gently wave your hand over the opening of a container and direct the fumes toward your nose. Do not inhale the fumes directly from the container.

25. Use only those chemicals needed in the activity. Keep all lids closed when a chemical is not being used. Notify your teacher whenever chemicals are spilled.

26. Dispose of all chemicals as instructed by your teacher. To avoid contamination, never return chemicals to their original containers.

27. Be extra careful when working with acids or bases. Pour such chemicals over the sink, not over your workbench.

28. When diluting an acid, pour the acid into water. Never pour water into an acid.

29. Immediately rinse with water any acids that get on your skin or clothing. Then notify your teacher of any acid spill.

Using Glassware Safely

30. Never force glass tubing into a rubber stopper. A turning motion and lubricant will be helpful when inserting glass tubing into rubber stoppers or rubber tubing. Your teacher will demonstrate the proper way to insert glass tubing.

31. Never heat glassware that is not thoroughly dry. Use a wire screen to protect glassware from any flame.

32. Keep in mind that hot glassware will not appear hot. Never pick up glassware without first checking to see if it is hot. See #22.

33. If you are instructed to cut glass tubing, fire-polish the ends immediately to remove sharp edges.

34. Never use broken or chipped glassware. If glassware breaks, notify your teacher and dispose of the glassware in the proper trash container.

35. Never eat or drink from laboratory glassware. Thoroughly clean glassware before putting it away.

Using Sharp Instruments

36. Handle scalpels or razor blades with extreme care. Never cut material toward you; cut away from you.

37. Immediately notify your teacher if you cut your skin when working in the laboratory.

Animal Safety

38. No experiments that will cause pain, discomfort, or harm to mammals, birds, reptiles, fishes, and amphibians should be done in the classroom or at home.

39. Animals should be handled only if necessary. If an animal is excited or frightened, pregnant, feeding, or with its young, special handling is required.

40. Your teacher will instruct you as to how to handle each animal species that may be brought into the classroom.

41. Clean your hands thoroughly after handling animals or the cage containing animals.

End-of-Experiment Rules

42. After an experiment has been completed, clean up your work area and return all equipment to its proper place.

43. Wash your hands after every experiment.

44. Turn off all burners before leaving the laboratory. Check that the gas line leading to the burner is off as well.

Glossary

Pronunciation Key

When difficult names or terms first appear in the text, they are respelled to aid pronunciation. A syllable in SMALL CAPITAL LETTERS receives the most stress. The key below lists the letters used for respelling. It includes examples of words using each sound and shows how the words would be respelled.

Symbol	Example	Respelling
a	hat	(hat)
ay	pay, late	(pay), (layt)
ah	star, hot	(stahr), (haht)
ai	air, dare	(air), (dair)
aw	law, all	(law), (awl)
eh	met	(meht)
ee	bee, eat	(bee), (eet)
er	learn, sir, fur	(lern), (ser), (fer)
ih	fit	(fiht)
igh	mile, sigh	(mighl), (sigh)
oh	no	(noh)
oi	soil, boy	(soil), (boi)
oo	root, tule	(root), (rool)
or	born, door	(born), (dor)
ow	plow, out	(plow), (owt)

Symbol	Example	Respelling
u	put, book	(put), (buk)
uh	fun	(fuhn)
yoo	few, use	(fyoo), (yooz)
ch	chill, reach	(chihl), (reech)
g	go, dig	(goh), (dihg)
j	jet, gently, bridge	(jeht), (JEHNT-lee), (brihj)
k	kite, cup	(kight), (kuhp)
ks	mix	(mihks)
kw	quick	(kwihk)
ng	bring	(brihng)
s	say, cent	(say), (sehnt)
sh	she, crash	(shee), (krash)
th	three	(three)
y	yet, onion	(yeht), (UHN-yuhn)
z	zip, always	(zihp), (AWL-wayz)
zh	treasure	(TREH-zher)

active transport: process in which a cell uses energy to transport a substance into or out of the cell

amino acid: building block of protein

asexual reproduction: reproduction requiring only one parent

ATP: substance in which cells store energy

autotroph (AWT-oh-trohf): organism that can produce its own food, primarily through photosynthesis

carbohydrate: energy-rich substance found in foods such as vegetables, cereal grains, and breads; sugars and starches

cell: basic unit of structure and function in living things

cell division: process in which one cell divides into two cells, each of which is identical to the original cell

cell membrane: thin, flexible envelope that surrounds a cell and allows passage of materials into and out of the cell

cell theory: theory that all living things are made of cells, that all cells come from other cells, and that the cell is the basic unit of structure and function in living things

cell wall: outermost boundary of plant cells that is made of cellulose

chloroplast: cell organelle containing chlorophyll that is involved in the process of photosynthesis

chromatin: threadlike coils of chromosomes

chromosome: rod-shaped cell structure that directs the activities of a cell and passes on the traits of a cell to new cells

compound: two or more elements chemically combined

cytoplasm: region between the cell membrane and the nucleus

diffusion (dih-FYOO-zhuhn): process by which substances move from a higher concentration of that substance to a lower concentration of that substance; primary method by which substances enter and leave the cell through the cell membrane

digestion: process by which food is broken down into simpler substances

division of labor: division of work among the different parts of an organism's body that keeps an organism alive

DNA: nucleic acid that stores the information needed to build proteins and carries genetic information about an organism

element: pure substances that cannot be separated into simpler substances by ordinary chemical processes

endoplasmic reticulum (en-doh-PLAZ-mic ri-TIHK-yuh-luhm): tubular passageways in the cell through which substances such as proteins are transported

enzyme: chemical substance that helps control chemical reactions

excretion: process of getting rid of waste materials

fat: substance that supplies the body with energy and also helps support and cushion the vital organs in the body

fermentation: energy-releasing process that does not require oxygen; less efficient than respiration

heterotroph (HED-uhr-oh-trohf): organism that cannot make its own food

homeostasis (hoh-mee-oh-STAY-sihs): ability of an organism to keep conditions inside its body the same, even though conditions in its external environment change

ingestion: taking in food; eating

life span: maximum length of time an organism can be expected to live

lysosome (LIGH-suh-sohm): small, round structure in a cell involved in the digestive activities of the cell

metabolism (muh-TAB-uh-lih-zuhm): sum total of all chemical activities in an organism

mitochondria (might-uh-KAHN-dree-uh): powerhouses of the cell in which cellular respiration occurs

mitosis (migh-TOH-sihs): process in which the nucleus of a cell divides into two nuclei and the formation of two new daughter cells begins

nucleic acid: large, organic compound that stores information that helps the body make the proteins it needs; DNA or RNA

nucleus (NOO-klee-uhs): cell structure that directs all the activities of the cell

oil: energy-rich compound that is liquid at room temperature

organ: group of different tissues working together; the third level of organization in an organism

organ system: group of organs that work together to perform a specific function for the organism; the fourth level of organization in an organism

organelle: "tiny organs" that make up a cell

organic compound: compound found in living things that contains the element carbon

osmosis (ahs-MOH-sihs): term given for the diffusion of water through a membrane

photosynthesis: process by which organisms use energy from sunlight to make their own food

protein: substance used to build and repair cells; made up of amino acids

respiration: process in which simple food substances such as glucose are broken down and the energy they contain is released

response: some action or movement of an organism brought on by a stimulus

ribosome: protein-making site of the cell

RNA: nucleic acid that "reads" the genetic information carried by DNA and guides the protein-making process

selectively permeable: membrane that allows certain materials to pass through, but restricts other materials from passing through

sexual reproduction: reproduction usually involving two parents in which a sex cell from each parent unite to form offspring

spontaneous generation: hypothesis that states that life can spring from nonliving matter

stimulus: signal to which an organism reacts; change in the environment

stomata (STOH-muh-tuh): openings in the lower surface of the epidermis in a green plant that allows gases to enter and leave the plant's cells

tissue: group of similar cells that perform a special function in an organism; the second level of organization in an organism

vacuole (VA-kyoo-wohl): large, round sac in the cytoplasm of a cell that stores water, food, enzymes, and other materials

Index

Credits

Cover Background: Ken Karp
Photo Research: Natalie Goldstein
Contributing Artists: Illustrations: Warren Budd Assoc. Ltd., Anni Matsick/Cornell & McCarthy, Art Representative; Gary Philips/Gwen Goldstein, Art Representative; Raymond Smith, Mel Greifinger, and Martinu Schneegass. Charts and graphs: Function Thru Form and Don Martinetti.

Photographs: 4 top and bottom left: Robert & Linda Mitchell Photography; bottom right: Dwight Kuhn Photography; 5 top: D. Cavagnaro/DRK Photo; bottom left: John Cancalosi/Tom Stack & Associates; bottom right: Robert & Linda Mitchell Photography; 6 top: Lefever/Grushow/Grant Heilman Photography; center: Index Stock Photography, Inc.; bottom: Rex Joseph; 8 top: Dr. Tony Brain & David Parker/Science Photo Library/Photo Researchers, Inc.; bottom: James D. Watt/Animals Animals/Earth Scenes; 9 Daemmrich/Uniphoto; 10 and 11 Dan McCoy/Rainbow; 12 Roger Ressmeyer/Starlight; 14 top: NOAO/Science Photo Library/Photo Researchers, Inc.; bottom: NASA; 16 top: Sidney Fox/Visuals Unlimited; bottom: Lynn Margulis/Dept. of Botany/University of Massachusetts; 17 top: William E. Ferguson; bottom: Breck P. Kent; 18 top: David M. Phillips/Visuals Unlimited; bottom: Dan McCoy/Rainbow; 19 left: Wayne Lynch/DRK Photo; center: Kjell B. Sandved; top right: Cabisco/Visuals Unlimited; bottom right: Robert & Linda Mitchell Photography; 21 top: D. Cavagnaro/DRK Photo; bottom left: Frans Lanting/Minden Pictures, Inc.; bottom right: David M. Phillips/Visuals Unlimited; 22 left: Chris Newbert/Bruce Coleman, Inc.; center: Hans and Judy Beste/Animals Animals/Earth Scenes; right: Dwight Kuhn Photography; 23 left: Robert & Linda Mitchell Photography; right: Dwight Kuhn Photography; 24 top: Frans Lanting/Minden Pictures, Inc.; bottom: Ocean Images, Inc./Image Bank; 25 top: Robert & Linda Mitchell Photography; center: G. I. Bernard/Animals Animals/Earth Scenes; bottom: Tom Bean/DRK Photo; 26 Merlin D. Tuttle, Bat Conservation International; 27 Wolfgang Kaehler; 28 top left: Jim Brandenburg/Minden Pictures, Inc.; top right and bottom: Robert & Linda Mitchell Photography; 29 left: Larry Ulrich/DRK Photo; right: Joe McDonald/Visuals Unlimited; 30 top left: Zig Leszczynski/Animals Animals/Earth Scenes; top center: Breck P. Kent/Animals Animals/Earth Scenes; top right: Robert & Linda Mitchell Photography; bottom: Frans Lanting/Minden Pictures, Inc.; 31 Frans Lanting/Minden Pictures, Inc.; 32 left: John Cancalosi/Tom Stack & Associates; right: Robert & Linda Mitchell Photography; 35 left: Robert & Linda Mitchell Photography; right: William E. Ferguson; 36 Wayne Lankinen/DRK Photo; 37 top: Dr. Tony Brain/Science Photo Library/Photo Researchers, Inc.; bottom: Will & Deni McIntyre/Photo Researchers, Inc.; 41 Dwight Kuhn Photography; 42 and 43 CNRI/Science Photo Library/Photo Researchers, Inc.; 44 Leonard Lessin/Peter Arnold, Inc.; 45 left: Dr. Jean Lorre/Science Photo Library/Photo Researchers, Inc.; right: David M. Phillips/Visuals Unlimited; 46 Domenico Ruzza/Envision; 49 top: Lawrence Livermore Laboratory/Science Photo Library/Photo Researchers, Inc.; center: M. Abbey/Photo Researchers, Inc.; bottom: K. G. Murti/Visuals Unlimited; 50 top: David M. Phillips/Visuals Unlimited; bottom: Rod Planck/Tom Stack & Associates; 51 K. G. Murti/Visuals Unlimited; 52 CNRI/Science Photo Library/Photo Researchers, Inc.; 53 top: W. Burgdorfer, Rocky Mt. Labs. Computer Enhanced by Pix*Elation/Fran Heyl Associates; bottom: BPS/Tom Stack & Associates; 54 top: K. G. Murti/Visuals Unlimited; bottom Dr. Jeremy Burgess/Science Photo Library/Photo Researchers, Inc.; 55 CNRI/Science Photo Library/Photo Researchers, Inc.; 58 Ken Karp; 62 and 63 © Lennart Nilsson, THE INCREDIBLE MACHINE, National Geographic Society; 67 top: M. Sheetz, University of Connecticut Health Center/J. Cell Biology; bottom left: Breck P. Kent/Animals Animals/Earth Scenes; bottom right: Kevin Collins/Visuals Unlimited; 69 M. Abbey/Visuals Unlimited; 70 Chris Baker/Tony Stone Worldwide/Chicago Ltd.; 71 Nina Lampen/Phototake; 72 top: Science VU/Visuals Unlimited; bottom: D. T. Woodrum and R. W. Linck.; 75 Nick Nicholson/Image Bank; 76 Carolina Biological Supply Company; 79 John D. Cunningham/Visuals Unlimited; 80 and 81 Charles C. Place/Image Bank; 82 top: Obremski/Image Bank; bottom left: Garry Gay/Image Bank; bottom right: Nino Mascardi/Image Bank; 83 Dr. Jeremy Burgess/Science Photo Library/Photo Researchers, Inc.; 84 David M. Phillips/Visuals Unlimited; 85 Dr. Jeremy Burgess/Science Photo Library/Photo Researchers, Inc.; 86 Robert Frerck/Odyssey Productions; 87 top: Robert & Linda Mitchell Photography; center: Stephen Dalton/NHPA/Animals Animals/Earth Scenes; bottom: Bill Varie/Image Bank; 88 CNRI/Science Photo Library/Photo Researchers, Inc.; 89 top: © Lennart Nilsson, THE BODY VICTORIOUS, Dell Publishing Company; bottom: © Lennart Nilsson, A CHILD IS BORN, Dell Publishing Company; 90 top: Ron Sherman/Tony Stone Worldwide/Chicago Ltd.; bottom: Paul J. Sutton/Duomo Photography, Inc.; 91 left: Dr. Robert Hessler, SIO/Woods Hole Oceanographic Institution; right: J. Frederick Grassle/Woods Hole Oceanographic Institution; 96 Kimberly Butler; 97 left: Tony Bucci © 1990 Discover Magazine; right: Journal of Comparative Neurology, F. Nottebohm, T. M. Stokes, C. M. Leonard, copyright 1976. Reprinted by permission of Wiley-Liss, a division of John Wiley and Sons, Inc.; 99 top: John Lei/Stock Boston, Inc.; bottom: Movie Still Archives; 100 New York Public Library; 104 Chris Baker/Tony Stone Worldwide; 117 Robert & Linda Mitchell Photography; 119 Rod Planck/Tom Stack & Associates